INDUSTRIAL AND SPECIALTY PAPERS

INDUSTRIAL AND SPECIALTY PAPERS

Volume I–Technology

Prepared by a Staff of Specialists
Under the Editorship of

Robert H. Mosher
Director, New Business Development
Kimberly-Clark Corporation

and

Dale S. Davis
Professor Emeritus, Pulp and Paper Technology
University of Alabama

CHEMICAL PUBLISHING COMPANY, INC.
New York   1968

Industrial and Specialty Papers, Volume 1, Technology

ISBN: 978-0-8206-0166-3

Chemical Publishing Company:
www.chemical-publishing.com
www.chemicalpublishing.net

First edition:

© **Chemical Publishing Company, Inc.** – New York, 1968

Second Impression:

Chemical Publishing Company, Inc. - 2013

Printed in the United States of America

CONTRIBUTING AUTHORS

Maurice C. Beren, President, The Pyrotex Leather Co., Leominster, Mass.

Thomas W. Busch, Vice President, Research and Development, Appleton Coated Paper Company, Appleton, Wisc.

William D. Hedges, Vice President, Research and Development, Columbus Coated Fabrics Co., Columbus, Ohio (Division of The Borden Company).

Bert C. Miller, formerly President, Bert C. Miller Inc.

Robert H. Mosher, Director, New Business Development, Kimberly-Clark Corporation, Neenah, Wisc.

Ralph T. Nazarro, formerly Technical Director, Texon Inc.

Percy E. Pierce, Assistant Manager, Fundamental Research, The Glidden Co., Cleveland, Ohio.

Paul H. Yoder, Vice President, Pierce & Stevens Chemical Corporation, Kimberton, Pa.

Foreword

The publication of *Specialty Papers* in 1950, followed by *The Technology of Coated and Processed Papers* in 1952, presented information on these subjects in book form for the first time in the English language. The paper industry is a broad and complex one and this is especially true of paper converting. No one man or organization is in a position to cover completely the many facets of specialty papers. Editor Robert H. Mosher therefore chose a group of well-qualified men, each a specialist in his own field, to prepare chapters on specific subjects. The results were two informative volumes that have served as general reference works.

Since the appearance of these two volumes, substantial advances have been made in the technology of specialty papers. Numerous new synthetic polymers have reached commercial production and some of these are of value in coatings. Several of the longer-established materials have been greatly reduced in price, offering new opportunities in specialty papers. At the same time, the availability of unsupported plastic film in large quantities at relatively low prices has drastically affected some markets for coated papers. Growth of metal foils has also offered both an opportunity and a challenge. Advances in machinery for preparing and using coated papers have kept pace with the new materials. This combination of new and improved materials and machinery has made possible an ever-expanding opportunity for specialty papers.

To bring this information up to date, Mr. Mosher has again obtained the aid of specialists to handle capably each of the many phases of the industry. Their efforts are combined in *Industrial and Specialty Papers—Their Technology, Manufacture, and Use* in three volumes.

These comprehensive books should serve as valuable references to a wide audience. While they cover the technology of specialty papers, the language of the chapters is not highly technical but is written along practical lines for more general understanding. These volumes should be helpful to the student as a text on this varied and relatively complex subject. They should be a source of information and background to the marketing man. To the laboratory researcher, they should serve as a reference and source of ideas for new products. Others may find them of general interest.

The editors and their authors are to be congratulated on bringing up to date the extensive information in this field.

New York, N.Y.   Wm. H. Aiken
August 1967   *Vice President, Research,*
                                                    Union-Camp Corporation
                                          President, Technical Association
                                          of the Pulp and Paper Industry

Preface

The paper industry represents a dynamic and ever expanding technology. Probably in no other areas have there been more developments in products and attendant techniques and machinery than in the fields of specialty papers and the allied coated and processed papers, particularly since 1950.

This period has seen the explosion in the office copying field, the evolving requirements of electronic data processing equipment, the insatiable demand for business forms, and tremendous advances in the graphic arts industry.

These developments have made necessary a complete revision of *Specialty Papers* (1950) and of *Technology of Coated and Processed Papers* (1952) to bring them up to date. As before, the purpose of this new series is to assemble, digest, and concentrate the best available information on each subject and to interpret this knowledge in the light of sound operating practice. Together with the host of carefully selected entries in the bibliographies, the series should serve as an auxiliary text to the standard volumes used in the leading schools of pulp and paper technology and as an authoritative technical and marketing reference work.

To attain these objectives properly, the editors have closely examined the earlier books in the light of their own experiences and those of their associates. Several chapters required only minor revision to bring the descriptions of certain techniques into conformity with existing improved procedures. The great majority required complete rewriting to achieve adequate coverage of technology and equipment that were entirely unknown fifteen years ago. Several new areas that were in their infancy at that time have been added.

As far as possible the editors have depended upon the original authors to revise and rewrite in their fields of interest and experience. Some have passed on, however, and several have left the industry. In choosing replacements, as well as the experts to cover new subjects, the editors again sought practical, successful men, rather than mere theoreticians, who had contributed extensively to recent developments and could write about them in a lucid manner. Sound, demonstrated practice has been the approach, yet adequate theory has been provided where it will shed light on the techniques used in this particular phase of the paper industry. The present work is definitely *not* the result of an academic and unimaginative search of the literature.

Many recent references appear where they support the wide experience of the authors. The reader will notice a certain amount of overlapping in the coverage of some of the topics—an intentional feature that reflects the inherent nature of the industry. There has been a deliberate orientation toward the end use and marketing aspects of each product, in comparison with the earlier, more limited technical approach.

The editors wish to thank the thirty-seven authors who deemed it worthwhile to take time from their busy schedules to present current practice authoritatively. Thanks are due a number of industrial concerns for permission to use photographs and tables not available elsewhere. The efforts of Mrs. Bairnsfather and others of the staff of the Hayden and Engineering Libraries of Massachusetts Institute of Technology aided considerably, as did the painstaking work of Mrs. Dixie Merson and Marguerite S. Davis in reading proof.

August 1967   Robert H. Mosher
                                                  Neenah, Wisconsin

                                                    Dale S. Davis
                                                Bailey Island, Maine

Contents

chapter 1

Introduction

R. H. MOSHER

EARLY HISTORY OF SPECIALTY PAPERS

Modern development, techniques, and materials of specialty papers can best be highlighted against a background of early history.

Prior to 1800, when paper was produced by hand and limited to single sheets, printing and coating operations were simple and as primitive as the paper-making technique.

This hand-made paper, which was produced in small volume in Europe beginning in the 12th century, was used from the earliest days of its development as a base for printing. All books and documents were of course hand written or printed prior to the development in 1456 of movable block type by Johann Gutenberg at Mainz, Germany. Previously, the use of woodcuts or engravings was the only method of commercial printing. The early inks were formulated from carbon black and vegetable oil and later from carbon black and mixtures of vegetable and mineral oils. Colored inks were made with vermilion, cinnabar, indigo, cobalt, and copper oxide as the pigmenting or coloring medium.

The printing of the surface of hand-made paper was a problem with the shallow wood cuts and even shallower wood engravings available at that time, because the surface of the paper was relatively coarse. This difficulty was overcome to a certain extent by burnishing the surface of each sheet with some sort of smooth, glossy stone, but the process was slow and cumbersome. About 1540, the glazing hammer was developed; this improvement caused a feud between the stone glazers and hammer glazers that lasted for more than 150 years, but as in most cases, the mechanically-operated device finally won out. This process in turn was superseded about 1720 when the Dutch developed a method of calendering the sheets with a pair of polished copper rolls, a technique that led to the invention of the sheet plating machine.

The main object of all these processes was to smooth down the surface of the paper so that fine engravings and letter press block printing could be produced with the paper available at the time. The printed paper was used for its traditional pictorial value and could not be considered as a part of the specialty paper industry as we think of it today.

The real start of the paper-converting industry began with the development of wallpaper, or paper "hangings" as they were then termed. This development must apparently be credited to the Chinese, who also invented paper, but it was brought to the western world about 1620 when paper hangings were first made in France. The sheet was produced in its earliest stages by block printing on plain paper, but artisans soon felt the need for some sort of "grounding" or base costing that would be capable of holding up the color and preserving the brilliancy of its hues. A coating was developed from glue and pigments and this was applied with a brush to supply a "ground" or "color" for the printing process. Later, someone discovered that the coated surface could be smoothed mechanically and used as a base for other types of printing operations.[3] This was the first purely decorative application of paper, other than as an artists' medium, and marked the beginning of the specialty paper industry, even though the paper was still only available in sheet form.

MODERN DEVELOPMENTS

About 1800, the basic invention that changed the paper-making industry appeared—a machine for the production of paper in a continuous sheet. The first machine, invented by Louis Robert in 1798, was developed by Henry and Sealy Fourdrinier and involved the same fundamentals that had been in use for 2000 years. The machine, which bears the name of the two brothers who lost a fortune in its exploitation, was based on an endless wire screen that allowed the water to run through and form a felt or sheet of paper on its surface. This sheet was continuously stripped off the end of the wire and either cut into strips, which were hung up to dry, or wound up in a partially-dried state and later redried. A second type of paper-making unit, the cylinder machine, was independently developed in 1806. It contained a wire covered cylinder that rotated in a vat of stock, with the water again running through the screen and the fiber mat remaining on the surface. The two types were introduced to this country in 1816 and 1827 respectively, and soon both were in production. Machines of the same type, although modified, enlarged, and speeded up, are still used today, although the early papermakers would scarcely believe their eyes if they could see them in their present form.[4,5,6]

The availability of paper in continuous form instead of sheets soon led to the development of web calenders for smoothing the paper to produce good printing surfaces, but it was not until the middle 1800's that equipment was developed for coating the paper by means of a continuous operation. The use of coated papers had been increasing, and coated labels, fancy papers, and wallpapers were gradually attaining reasonable volume. Gravure printing that employed copper plates had been introduced about 1650 with the mezzotint process, and lithography was developed by Alois Senefelder in the early 1800's. Coated and printed playing cards were developed prior to 1500. All these products and processes had previously been limited by the lack of continuous machinery for their production.[2]

One of the first companies formed for the express purpose of coating fancy and box papers was the firm of Alois Dessauer, which was set up in 1810. By 1865 it had grown so that the employment had risen to 300 and a sizable volume of paper was being handled. The continuous coating process was brought into Germany in 1866 at the plant of E. Kretzschmar in Dresden by a Frenchman named Möglin. A similar machine appeared in England at approximately the same time and in apparent independence of the French development.

The first machines faithfully reproduced hand brushing by the use of the so-called "sun and planets" smoothing unit in which the coating was smoothed out by brushes that revolved in a motion that imitated the manual operation. These machines had obvious disadvantages, however, and in 1874 a cam-driven oscillating-brush unit, similar to those used through the 1940's in the paper converting industry, was developed at the Kretzschmar plant. The evolution of the method of making paper in a continuous web and the subsequent adaption of the hand coating technique to the continuous machine-coating operation was the basis for the specialty paper industry of today. Continuous methods of printing, staining, embossing, brushing and otherwise decorating paper for the fancy and decorative trade followed in relatively short time.[1,2,4]

The development of the specialty paper industry in the United

States followed along the same lines, but with a time lag of several years subsequent to the European evolution. As noted previously, the first Fourdrinier paper-making machine was set up in this country in 1816 and the first cylinder machine in 1827. The wallpaper industry was a going business at that time as the first factories had been set up in 1775, and by 1795 mills were producing this type of decorative sheet in Pennsylvania, New Jersey, and Massachusetts. The early papers were made in sheets about 30 inches long. [3] As early as 1824, a wallpaper with a coated and glazed background for the usual block printing was manufactured and the development was rapidly picked up by producers of fancy papers. Aside from wallpapers, two other growing industries—label printing and paper box manufacuring—were rapidly perfecting the art of coloring, glazing, and printing paper for both decorative and traditional applications.

In 1839, the firm of Pollack and Doty was formed in Philadelphia for the express purpose of producing coated label papers. These papers were coated and plated and then printed in the form of labels for print works, cotton mills, and proprietary medicines. According to Wheelwright, [3] "The paper was coated sheet by sheet, the color being brushed on it by girls whose aprons were covered with blotches of all sorts of colors; the sheets were then hung on laths which rested

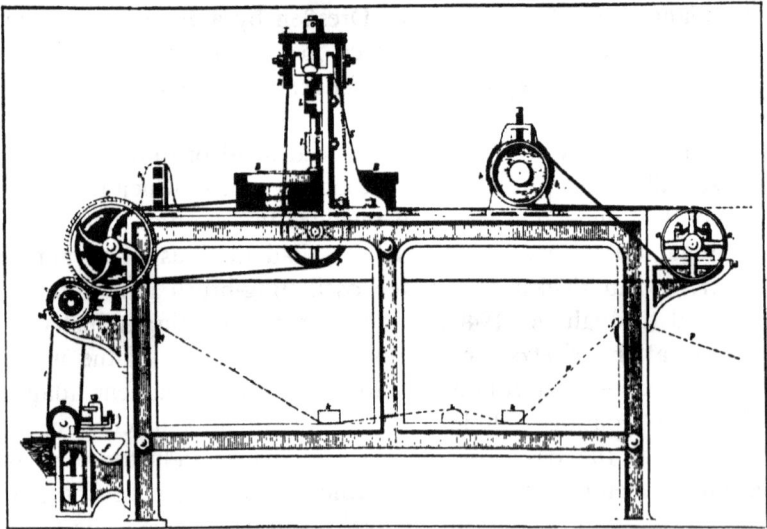

Fig. 1-1. Wall paper coating — about 1860 (*Courtesy of Paper Museum of M.I.T.*).

on wooden racks, where they remained overnight—sometimes longer —to dry."

The industry showed a slow but steady development and growth with the formation of new converting companies at Albany, N.Y. and Nashua, N.H., both of which did large and prosperous business. In 1845, the firm of Doty and Bergen was started in New York; it later changed to Doty and Scrimgour and the plant moved to Reading, Pa. In 1846, the firm of J. & L. Dejonge was formed at Staten Island and was set up to produce glazed fancy papers. This firm in 1852 bought a coating machine from the firm of John Waldron Co. (which had been producing coating machines since 1832 for the wallpaper industry) for the purpose of making glazed paper on a continuous basis—an art that was to revolutionize this phase of the industry. The machine, which was of the "sun and planet" design, was bought for $225.00.[3] The web of paper was coated and festoon-dried before being rolled up. The rolls were then friction-calendered or flint-glazed, and sold, either plain or embossed, for labels and box coverings. In time, rotary brushes for applying color and oscillating brushes for smoothing the surface gradually supplemented and replaced the older designs, but the flinting and friction-calendering operations remained practically the same as when they were first developed.

In 1875, a new field of application of coated papers was developed when Theodore Low Devinne conceived the idea of using such paper for printing purposes where good color reproduction was desired. [3] Up to this time, paper had only been coated on one side, but for such an application two-side coating was a necessity. Charles M. Gage, who was manufacturing one-side coated and glazed paper in Springfield, Mass., accepted the challenge and produced 100 reams as an initial order. The resulting paper was printed from fine wood engravings and the results were very satisfactory. The idea was then temporarily dropped until 1881 when Gage, then associated with the S.D. Warren Company at Cumberland Mills, Me., again accepted the challenge and produced another quantity of the paper. The resulting printing work was so superior to the supercalendered paper then in use for gravure printing that soon afterward, when the photoengraving process was developed, it became the standard paper for fine halftone printing.

NEW TECHNIQUES

In more recent years many further developments have been forth-coming in this industry. Improved adhesives that employ starch, casein, soya protein, and synthetic resins and rubbers have been advanced to take the place of the glue that was used in early work. Better and more light-fast dyes and pigments, as well as specialized chemical additives, have been constantly introduced and have raised the standards of the resulting coatings.

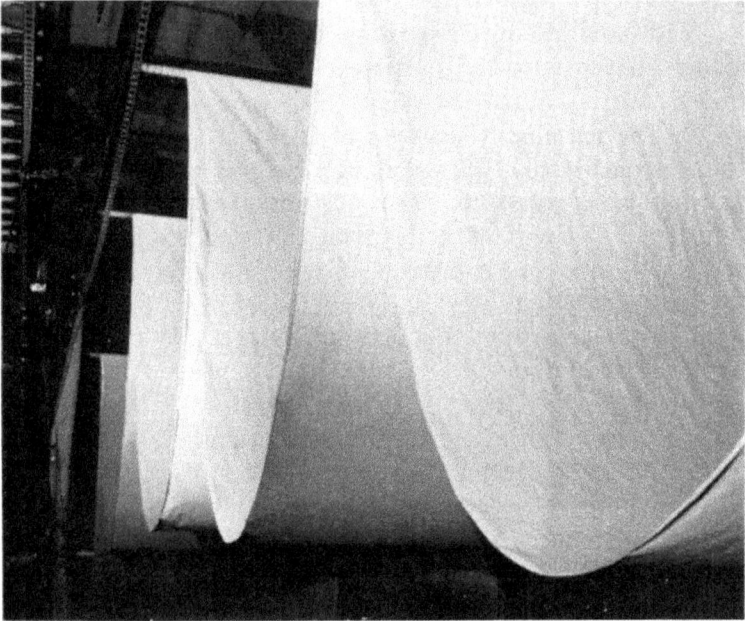

Fig. 1-2.   View inside festoon drier (*Courtesy of Watervliet Paper Co.*).

Dull- or matte-finished papers, gummed papers, mica-coated papers, and finally, the varnished and lacquered or plastic-coated papers, were developed. New methods of coating, which eliminate the need of the brush coater and based on roll or spray applications, and the use of an air knife, rolls, or various types of doctor instead of brushes for smoothing the coated surface, have been introduced and brought to a high degree of efficiency. The old festoon lines have been modernized and gradually replaced by enclosed tunnel driers that increase the speed and efficiency of drying. All these developments have tended to improve the variety and quality and to reduce the unit cost of coated papers.

By 1950 the paper converting industry had only begun to emerge from its self-sufficient and self-imposed shell. Specialty products were still being manufactured in a series of small unit processes by operators with narrow interests. They were truly converters in that the wide variety of base papers, boards, films and foils were made elsewhere by other organizations and shipped to them for processing and finishing. Major exceptions were those that made machine-coated publication printing papers directly on the paper machine. This group, however, was a closely-knit club, because its highly specialized technology required heavy capital investment in large paper machines.

High grade printing papers as a rule were still manufactured by means of off-the-machine conversion techniques. The development

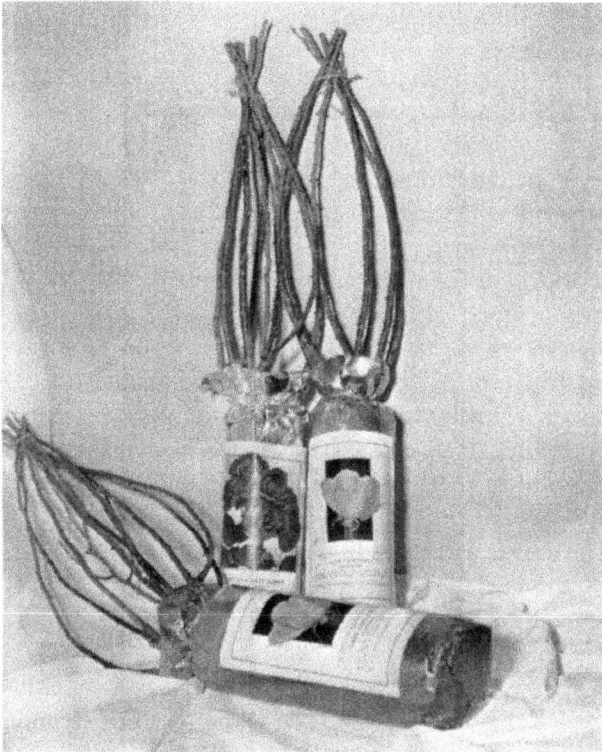

Fig. 1-3. Asphalt laminated aluminum foil nursery wraps (*Courtesy of Thilmany Pulp & Paper Co.*).

and commercialization of high-speed blade, off-the-machine coaters later enabled not only producers of routine publication paper to upgrade their products, but also opened the door to the smaller manufacturer who could pass the production of several paper machines through one piece of high speed, off-the-machine coating equipment. By 1960 numerous versions of such coaters became readily available from several of the large machine manufacturers who were also in a position to recommend coating formulations and know-how where these were necessary. The expansion in the commercial printing industry and the growth of markets for specialty coated paper were reflected directly by expansion of the coated paper industry.

A second major factor and one that altered the profile of the industry was the trend toward mergers of the smaller companies and the acquisition of the smaller by the larger ones. Over a period of years this tendency has eliminated most of the specialty converters as corporate entities, but in the process it has introduced the knowledge and technology of these highly specialized organizations into fertile ground for growth of the large corporation. This situation quickly made possible long-needed expansion of the specialty organization through the infusion of readily available capital and the broad technical and engineering capabilities already present in the large organizations. The market research and efficient marketing and sales efforts of the combined organizations have made possible the expansion of the industry in terms of total production as well as the broader range of markets serviced.

A list of the specific areas where the converting industry has expanded and where other industries are now invading the paper field would touch practically any of us in our daily lives. Papers of distinctive appearance or unique characteristics are constantly sought by the package designers and others who create the appealing products and packages that are demanded by a quality-minded purchasing public. The more useful and specially designed papers have often replaced cloth, coated fabrics, leather, or wood. The market for papers with unusual properties is most lucrative.[8] A look at the diversified utility of paper and paper-based products is persuasive evidence that they represent a rare combination of potential characteristics whose commercial acceptance is the result of a similar pattern in each instance—the most practical way to achieve a group of difficult specifications, unusual features, or special properties.

SYNTHETIC MATERIALS

Another feature that has revolutionized the paper-converting industry is the development and practical exploitation of synthetic resins, rubbers, and plastics. Nitrocellulose lacquers were found to be ideal coatings for paper and made possible decorative and functional effects that could not be obtained with the previously available substances. Other cellulose derivatives and the synthetic resins and natural and synthetic rubbers, which were developed with startling speed, opened new fields where paper was a definite winner, or at least a very close competitor instead of a poor substitue for more expensive materials. As the new plastics and chemicals were generally insoluble in water, an entirely new application technology had to be worked out. However, this time the development was mainly in the hands

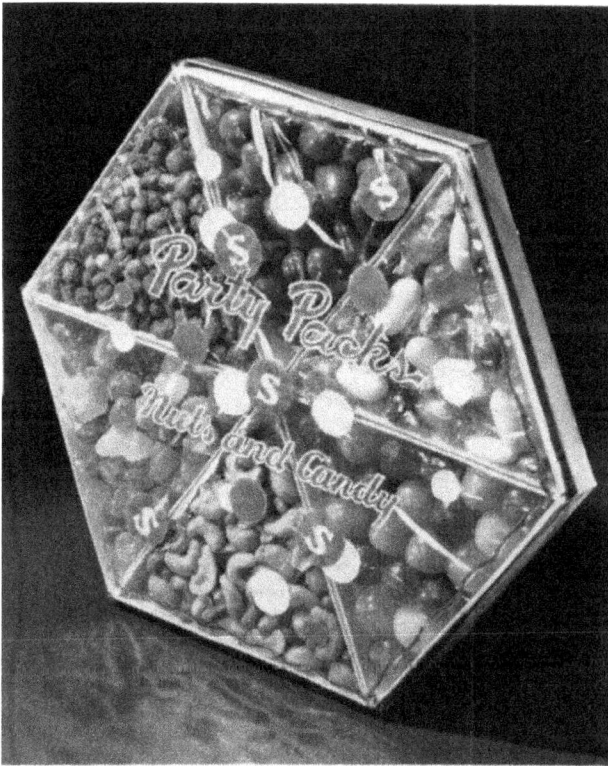

Fig. 1-4. Typical fancy wrapped candy package (*Courtesy of Sylvania Division of American Viscose Corp.*).

of the machine builders and the large chemical companies that were responsible for the discovery and exploitation of the plastics. These large and relatively wealthy companies, instead of the paper companies with their more limited facilities, evaluated the new products and in cooperation with progressive paper converters soon developed new combinations of paper and plastic as well as economical methods for their manufacture.

Because the same resinous coating could often be applied by means of several different techniques, similar yet competitive products could frequently be obtained. Solvent coatings were the first developed and exploited, closely followed by the hot melts. As plastics technology and chemical process development advanced to the stage where emulsion polymerization became a commercial reality, water dispersion coatings were produced. These made possible the application of plastic films from an aqueous system.

The most recent advance in the field is the development of paste dispersion as well as extrusion coatings, and with them the possibility of laying down thick films with a minimum effort. The use of encapsulation techniques offers an additional approach to the application of coatings with specialized purposes. The development of various plastic films and the possibility of utilizing metal foils and vacuum metallizing techniques for decorative and functional products now permit specialty converters to produce sheets with appearances and properties that were not even imagined two decades ago. The products of the paper converter do not merely compete with wood, metals, glass, and other materials, but have often replaced them in their own fields.

STATE OF THE INDUSTRY

Considerable technical and marketing information is available in the plastics-treating field inasmuch as every machine and chemical company that attempts to merchandise its products has available technically trained sales and service men and maintains customer-service laboratories for the solution of specific converter problems. The result has been that the basic information necessary to enter almost any phase of the paper converting industry is now available in many places, but the "know-how" on specific sheets is still mainly, as it should be, in the hands of the individual manufacturers. The purpose of this series is to concentrate the available information on the various phases of the industry, to temper this knowledge with operating "know-how", and to condense the subject so that along with the listed references these volumes can be used as a starting point to study any specific phase of the speciality converting field.

The many available processes, along with an extremely wide range of suitable raw materials, make possible the manufacture of a similar (or even identical) product by different processes from different raw materials. For example, a greaseproof decorative wrapper can be made from a saturated or plasticized base sheet that can be coated, printed, laminated, and calendered or embossed. If it has a pigment coating, it may have been made on a brush, doctor, roll, spray or air-knife coating unit. If it is a plastic-type sheet, any one of the several dozen base resins or rubbers could have been used in the coating formulation and the coating could have been laid down from a solvent, water dispersion, hot melt, organosol, plastisol, or extrusion applicator. The printing may have been done on an aniline press, or by the gravure, letterpress, electrostatic, or surface technique. If the paper is embossed, it may have been handled on a geared, flat-back, or meshed-roll machine. The costs and rate of production vary widely with differing equipment or even with the same type of machine in different mills. The specialty-paper industry is fairly unique in the variety of its manufacturing operations, and every mill unit has developed its own modification of the usual practice.

To date this growth seems to show that the future will continue to offer great promise to the specialty paper and converting industry. According to Macdonald,[7] however, great resourcefulness on the part of the paper-use promoters will be required. The enormously expanding field of plastics products can become predominantly a branch of the chemical or the paper industry depending upon which makes the greater product development and promotional efforts. The opportunity for development and promotion is further dependent on the quality and intensity of technical effort. Between these two industries is the converter who takes paper and additive materials and combines them to create products that have a specific market value.[7] The ultimate user buys the best product designs available at a given price and cares little about how, or of what, they are made. If the independent converter can direct his efforts along similar lines he will be able to create products to meet the insatiable demand of the consuming public.

Against the production of specialties can be contrasted the more standard pigment-coated papers of the glossy and matte type for the fancy paper field, and the enormous quantities of coated stock for the printing-paper industry. In these areas only a limited number of adhesives are available, relatively few standard pigments, and only a half-dozen methods of economic application. This part of the industry attempts to be as secretive with its coating formulations and technology as are the highly specialized portions.

REFERENCES

1. Carlton Ellis, *Printing Inks*, Reinhold, New York (1940).
2. August Weichelt, *Buntpapier-fabrication*, Carl Hofmann-Verlag für Papier-Zeitung, Berlin.
3. W. B. Wheelwright, "The Coating of Paper," *The Papermaker*, 1, 1-5 (February 1948).
4. Anon., *The History and Story of Paper Making*, Kalamazoo Vegetable Parchment Co. Roycrafters, East Aurora, N.Y.
5. Anon., *Paper: Pacemaker of Progress*, F. C. Huyck and Sons (C. Lynn Summer Co.), The Marchbanks Press (1946).
6. Canfield, "The Paper Industry," *Think* (May 1947).
7. R. G. Macdonald, "Postwar Possibilities of the Paper Industry," *Barrell's Paper Annual* (1944-1945), p. 3.
8. J. K. Speicher and F. K. Shankweider, "Creative Coatings—Lacquers for Papers," *The Papermaker*, 1, 7-11 (February 1948).

chapter 2

Theory of High Polymers and Their Application to the Paper-Converting Industry

RALPH T. NAZARRO†

Much scientific work has been published on the theory of the formation and reactions of high polymers, so that any attempt to condense this vast amount of data is bound to be lacking in many respects. From the standpoint of the chemist in the paper-converting industry, the criterion by which a given resin is judged is the usefulness of one or more of its properties. An understanding of the theory and chemistry of high polymers simplifies the evaluation of these properties and permits the practical chemist to utilize a given resin to the fullest extent.

The scope of this book does not permit a thorough discussion of the various physical and chemical properties of resinous systems, but general statements are supported by a generous number of literature references in which the original data can be found in detail.

POLYMERIZATION

Berzelius [1] first introduced the term "polymer" to differentiate the product or products of ordinary chemical reactions from those reaction products whose composition remained similar to that of the original substances. Early theories were unable to characterize the types of polymer, the possible mechanism of high-polymeric reactions, and the influence of functional groups. Polymers have been classified according to their formation mechanism and also according to the chemical groups they contain.

† Deceased.

The two general types of polymer are: *addition* or association polymers and *condensation* polymers. Addition polymers are built up by the addition reaction of their monomers to form a molecule of macromolecular weight. The formation of polystyrene, polybutadiene, polyvinyl acetate, and polyvinyl chloride, are examples of additive polymerization. The structure of such polymers is sometimes complicated by bridging and cross-linking of the chains, particularly in conjugated olefinic systems.

The polymerization of styrene, which forms a resin on heating, to polystyrene under the most favorable conditions may be written as follows[98]:

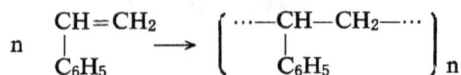

$$n \quad \begin{matrix} CH{=}CH_2 \\ | \\ C_6H_5 \end{matrix} \quad \longrightarrow \quad \left[\begin{matrix} \cdots{-}CH{-}CH_2{-}\cdots \\ | \\ C_6H_5 \end{matrix} \right]_n$$

Carothers and others [12, 20, 23, 67, 68, 70, 104] agreed that with allowance for the initiating structure and terminal groups this structure appeared to be the most plausible of those suggested. The original compound, e.g., styrene, which constitutes the recurring unit in a polymer, is called a monomer.

The desire to modify the properties of a given polymer led to the development of copolymerization. In this reaction, two or more monomers are allowed to polymerize together to form a copolymer. The macromolecule in this case consists of recurring units of the original monomer mixture. A copolymer has physical and chemical properties characteristically different from those of a mixture of polymers derived from the polymerization of the individual components of the copolymer. The synthetic rubber GR-S is a copolymer of styrene and butadiene. By varying the proportion of butadiene and styrene in the original monomeric mixture, a series of copolymers can be obtained. These range from soft, low tensile-strength rubbers of a high butadiene content to firm, high tensile-strength films with a high styrene concentration. Recent studies [169, 174] on addition polymerization have emphasized the connection between molecular structure and the physical properties of high polymers as an aid in designing special "custom-made" molecules. The comonomer, in such cases, is actually an auxiliary monomer and is employed in small amounts to increase the mutual attraction between the chains or to stiffen the individual chain molecules.

To the practical chemist, the success of this research would mean plastics of higher softening points with greater resistance to chemical change at higher temperatures. Copolymers of styrene with acrylonitrile, for example, have been prepared with softening points considerably higher than those of polystyrene. The introduction of the acrylonitrile distributes strong molecular attraction along the chain, which tends to lend improved resistance to deformation. The introduction of large side groups along the chain, e.g., naphthalene and phenanthrene, which decrease the internal flexibility of the macromolecule, also yields polymers and copolymers with superior thermal properties.

Addition polymers and copolymers can be considered as derivatives of the vinyl group. Among the vinyl polymers used with paper are polythene, polystyrene, polyisobutylene, polyalkyl acrylates, polyalkyl methacrylates, polyvinyl acetate, polyvinyl chloride, and polyvinyl ethers. The most available synthetic rubberlike polymers and copolymers are derived from, or are made with, the conjugated diolefin, butadiene. The synthetic rubbers used in paper converting include the copolymers of styrene–butadiene, acrylonitrile–butadiene, isoprene–isobutylene, and the polymer, polychloroprene.

The second general type of synthetic high polymer is the product of condensation reactions between at least two substances. A condensation polymer is formed through the elimination of water from the reactants.

The behavior of monofunctional compounds is the basis of many elementary organic preparations without resin formation, e.g., the formation of ethyl acetate by condensing acetic acid and ethyl alcohol. However, the condensation of polyfunctional compounds, such as ethylene glycol and maleic acid, leads to the formation of a long-chain polyester called an alkyd resin. If one of the reactants of the condensation mixture contains more than two functional groups, e.g., glycerol, then the resin formed may be a cross-linked polyester.

Cross-linking of the chain yields a polymer with better resistance to water and solvents and greater dimensional stability. The development of cross-linked three-dimensional structures can be better accomplished by the introduction of a third substance capable of copolymerizing with any residual unsaturation within the molecule. Styrene and cyclodienes have been added to polyesters containing unsaturation, e.g., an ethylene glycol–maleic acid reaction product, and the polymerization completed after a predetermined curing cycle.

The formation of a cross-linked polyester follows the general condensation reaction between a polyfunctional alcohol and a polyfunctional carboxylic acid, but after the condensation either the alcohol or the acid must contain some residual unsaturation capable of copolymerizing with a hydrocarbon vinyl derivative. The polyester alkyds have also been used with drying or oxidizable oils for paper coatings.

The formation of these condensation polymers involves only the carboxyl group and the hydroxyl group, but other functional groups also yield important and well-known condensation polymers. The condensation of urea and formaldehyde yields a hard polymer on dehydration of the methylol intermediates. Phenolics are produced by acid- or alkali-catalyzed condensation of phenols and formaldehyde in various concentration ratios. The formation of polyamides, from which the nylons are derived, involves the condensation of diamines and dicarboxylic acids. The ureas, phenolics, and condensation products of melamine and formaldehyde have been extensively investigated in the development of wet-strength papers. Condensation polymers from the reactions between rosin abietic acid and polyhydric alcohols such as pentaerythritol and glycerol have been employed in the sizing and coating of decorative papers.

Some resins derived from cellulose are not in any definite class of either the vinyl resins or the condensation polymers. They are referred to as cellulose ethers—e.g., methyl and ethyl cellulose and hydroxyethyl cellulose—or as cellulose esters—e.g., cellulose nitrate and cellulose acetate. All cellulose derivatives have been used in emulsion or solution coating of functional papers.

A similar situation exists with rubber derivatives such as chlorinated rubber and rubber hydrochloride. Their use is limited to the development of special oil- and chemical-resistant sheets by applying them in solution to paper by means of a conventional coating method.

Briefly, the addition polymers and copolymers can be considered as derivatives of the vinyl group, and form high polymers whose molecules are represented in the normal unstretched state as a mass of randomly coiled, irregularly entangled chains capable of slipping past one another without cleavage of the linear chain. The individual segment, therefore, lends flexibility and elasticity to the polymer system. In contrast, the fully-saturated condensation polymer has a rigid structure, the degree of rigidity depending on the number of functional groups of the original reactants, and also on any subsequent copolymerization that would give cross linking of the molecular chains.

Chemical Structure and Polymerization

The literature on the formation of high polymers from their respective monomers [19, 24, 69] is very extensive; unfortunately, some of the results are contradictory and unnecessarily complicated. Chemists agree that an addition polymerization is a chain reaction and that a given chain is terminated by collision with an ion or radical or by transferring energy to another molecule. Addition polymers are the characteristic products of olefinic compounds. Apparently, the reacitivity of olefins and, therefore, polymerization can be interpreted in terms of the electron density of the double bonds. The presence of atoms or groups of atoms that enhance the polarization or polarizability of the olefinic compound normally increases the ease with which this compound polymerizes. Asymmetrical substitution of olefins, in general, improves the polymerizability of a given compound, but substitution on both carbon atoms of the double bond reduces markedly the degree and rate of polymerization.

The vinyl derivatives represent the largest class of resins currently used in synthetic coatings for paper. The manner in which they are formed, particularly the initiating steps, is still in dispute, but is associated unquestionably with the activation requirements of the double bond. Each valence dash in the $C=C$ double bond represents a pair of electrons, according to the Lewis valence theory, and in the quantum mechanics concept each electron pair has a certain charge distribution which Schrödinger [45, 83, 99, 122] has represented mathematically by suitable eigenfunctions. The wave functions of the two electron pairs are distinguishable as eigenfunctions of the first order and second order, respectively.

The first-order eigenfunction corresponds to the normal single C—C bond and is of no significance in the initiation reaction. Polymerization is caused by the eigenfunctions of the second order or the "unsaturated electrons." Activation of the monomer, therefore, is the process of raising one or both of the unsaturated electrons to higher energy levels than the one that is normally characteristic of the molecule in its stable state.

In the absence of a catalyst, the activated state of the double bond of ethylene, for example, is probably highly polar:

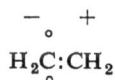

$$\overset{-}{\underset{\circ}{}} \quad \overset{+}{}$$
$$H_2C:CH_2$$

In this formula, electrons of the first order are designated by points and those of the second order by small degree circles.

Thus, the thermally initiated molecule polymerizes by collision of the active structure with another monomer molecule and continues to grow into long chains until deactivated by energy transfer or terminated by the influence of other ions or radicals.

Considerable difference of opinion and conflicting data exist on the initiation reaction via free radical formation.[58,70,82,84,89] Both free radical formation and activated double bonds may be responsible for the initiation reaction. Staudinger[100] postulated in his early work on polystyrene that the reaction proceeded through the formation of free terminal bonds or free radicals. However, several authors[69] have questioned the formation of free radicals and the part they play in polymerization reactions.[19]

Assuming that the molecule receives sufficient energy to reach its particular energy hump, then it may yield a radical that continues to activate other molecules and to propagate the process. The activation of styrene and its derivatives could conceivably be the result of both free radical and activated double bond response, thus:

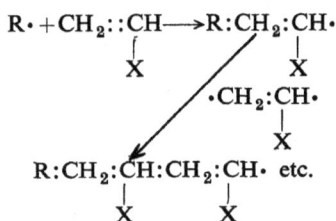

$$R\cdot + CH_2::CH \longrightarrow R:CH_2:CH\cdot$$
$$\underset{X}{|} \qquad \underset{X}{|}$$

$$\cdot CH_2:CH\cdot$$
$$\underset{X}{|}$$

$$R:CH_2:CH:CH_2:CH\cdot \text{ etc.}$$
$$\underset{X}{|} \quad \underset{X}{|}$$

$R\cdot$, a free radical, causes the electrons to shift and form a new free radical that propagates the process.

Compounds such as styrene that contain a negative group attached to the vinyl radical tend to form polymers in which the degree of polymerization is high. In general, asymmetric substitution in olefins, cycloolefins, and diolefins facilitates polymerization.[25,26,27,102]

Chloroprene, or 2-chlorobutadiene-1,3, polymerizes about 700 times as fast as isoprene, whereas bromoprene, or 2-bromobutadiene-1,3, polymerizes roughly 1000 times as rapidly as isoprene. Iodoprene, or 2-iodobutadiene-1,3, polymerizes completely in 48 hours at ordinary temperatures compared with about 130 hours for the polymerization of the bromo-compound, and about 10-25 days for the thermal polymerization of chloroprene at room temperatures.

Permanent polarization of a polymerizable compound gives rise to increased reactivity and the same is true of any contribution from resonating structures. A comparison of the polymerization rates of ethylene and vinyl chloride bears out the marked influence of substituent negative groups. Structures contributing to the activity of vinyl chloride have been illustrated as follows [123]:

$$\overset{-}{Cl}-\overset{+}{CH}=\overset{+}{CH_2} \qquad \overset{+}{Cl}=\overset{\frown}{CH}-\overset{\longleftarrow}{CH_2}$$

$$\underset{induction}{\longleftarrow} \qquad\qquad resonance$$

To a great extent, polymerization processes are activated, by ionic catalysts[57,93,101,102,115,124] such as $SnCl_4$, $AlCl_3$, BF_3, concentrated H_2SO_4, and $(C_6H_5CO)_2O_2$. The ability of the olefinic compound to release electrons is responsible for the electrophilic attack at the double bond, promoting an active complex that persists in its influence on other monomeric units and continues to cause electron shifts that propagate the reaction. Although ionic catalysis proceeds with difficulty when the availability of electrons at the double bond is inhibited by the presence of electron-deficient groups adjacent to the olefin bond, the general mechanism[74a,88] for ionic catalysis can be represented by

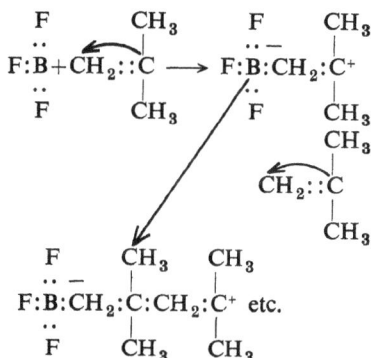

The peroxide-initiated polymerization of vinyl derivatives is generally believed to proceed by a free-radical chain mechanism, the elemental steps of which include activation, propagation, chain transfer, and termination.[2,29,93,115] The activated complex formed by the addition of initiating peroxide to the olefinic double bond has the character of a free radical and initiates a chain reaction.

Studies[7,11,60,61,85,86] have been made of the action of benzoyl peroxide in catalyzing the polymerization of vinyl monomers. The evidence generally points to the cleavage of the peroxide into two free phenyl or benzoyl radicals:

$$R-\overset{\overset{\displaystyle O}{\|}}{C}-O-O-\overset{\overset{\displaystyle O}{\|}}{C}-R \longrightarrow 2R-\overset{\overset{\displaystyle O}{\|}}{C}-O\cdot$$

$$R-\overset{\overset{\displaystyle O}{\|}}{C}-O\cdot \longrightarrow R\cdot + CO_2$$

Early reports [39] on the decomposition of peroxides indicate that the active fragments are hydrocarbon radicals, but recent investigations,[60,61,86] with diacid peroxides present analytical data concerning the polymeric products to substantiate the presence of some oxygen-initiating fragments.

The disposition of the resultant radicals depends on the reaction medium and the temperature of the reaction system. Apparently, the ease with which the hydrocarbon radical forms over that of the benzoyl radical is affected by the reaction solvent[78] and, to a greater extent, by the temperature. Decarboxylation can also occur in bulk polymerization in the absence of any reaction solvent.

Under given reaction conditions, the peroxide decomposes into more-active fragments and either the hydrocarbon radical, the acid radical, or both may combine with the monomer.

$$R-\overset{\overset{\displaystyle O}{\|}}{C}-O\cdot + CH_2 \!=\! CHX \longrightarrow R-\overset{\overset{\displaystyle O}{\|}}{C}-O-CH_2-\underset{\underset{\displaystyle X}{|}}{CH}\cdot$$

$$R\cdot + CH_2 \!=\! CHX \longrightarrow R-CH_2-\underset{\underset{\displaystyle X}{|}}{CH}\cdot$$

A new free radical is created that can, in turn, activate another molecule of monomer unit. This chain reaction continues until the free radical at the end of the growing chain is destroyed by some foreign molecule or by combination with another free radical. Thus, a benzoyl radical may terminate a chain as well as start one, but two growing polymer chains are more likely to collide and form a permanent chemical bond.

Methods for the Preparation of Vinyl Polymers

The four main procedures for polymerization of vinyl monomers are: Bulk polymerization, solution polymerization, suspension polymerization, emulsion polymerization.

Bulk Polymerization

Mass polymerization, as this method is frequently called, is capable of producing polymers of good optical clarity and generally superior electrical properties, but the degree of polymerization is comparatively low because of the difficulties in dissipating the heat of polymerization.

Several modifications of bulk polymerization have been devised for specific industrial applications.   The usual laboratory method is carried out in a series of suitable glass cylinders, rotating constantly at a given temperature and so constructed that representative samples can be removed[69] frequently for comparative studies of activation energies and polymerization rates.[44]

Most liquid polymerizations are carried out under normal pressure and from room temperature to above the boiling point of the monomers.   Polymers of ethylene were produced successfully by subjecting ethylene[14,31,55,66,69,121] to pressures up to 1200 atm and to temperatures of 100-300°C.   Styrene has been successfully polymerized thermally and catalytically in bulk processes in the laboratory and in industry.   For example, at 100°C under atmospheric pressure, styrene gradually becomes viscous as the polymerization progresses.   The resin is water-clear provided that precautions are taken to remove atmospheric oxygen from the polymerization container.

Solution Polymerization

The two facts that emerge in a study of solution polymerization are that the presence of a solvent always slows down the polymerization reaction and that at the same time the average molecular weight of the polymer is comparatively low.

Many authors[2,5,6,13,21,23,33,40,56,61,69,89,117,118,119] have written on the mechanism of polymerization of ethylene derivatives, particularly styrene, in a homogeneous liquid phase.   A nonsolvent for the polymer, such as methanol, may be present along with a good solvent such as benzene or toluene to precipitate the polymer as it is formed. In that event, molecular weight distribution of the final polymer can be controlled in two ways: First, by dissolving polymers of lowest molecular weight in a mixture of methanol and benzene and thus largely freeing the solution from polymers of high molecular weight; second, by precipitating a growing chain from solution as soon as its solubility has been exceeded because of molecular size.

Collision between a growing chain and an ion or radical of the solvent frequently results in a termination reaction.   Confirmation of chemical combination between chain and solvent can be found in the analysis of styrene polymerized in carbon tetrachloride.   The solvent in this case not only reduces the molecular weight of the styrene polymer by donating an atom to the growing polymer radical, forming a polymer molecule,[3,43,79,80] but also measurably alters the physical properties of the resulting products.[81]

F. R. Mayo[74] summarized the previous work on the polymerization of styrene in solvents, although the effects of solvent[95] on this polymerization were known for sometime, as indicated by Table 2-1.

TABLE 2-1

Effect of Solvents on the Polymerization of Styrene at 100°C *

Solvent	Styrene Concentration, %	Molecular Weight of Polymer
None	100.0	200,000
Toluene	50.8	180,000
Toluene	26.0	140,000
Toulene	12.0	86,000
Toluene	6.0	60,000
Toluéne	1.6	20,000
Carbon Tetrachloride	80.0	24,000
Carbon Tetrachloride	46.0	9,000
Carbon Tetrachloride	33.0	6,500

* Z. physik, Chem., A179, 36 (1937).

Obviously, the lower the concentration of the monomer in the solvent, the greater the probability of collision between the monomer particle and solvent. The result is an inevitable termination of growth of the polymeric chain. This particular effect is apparent regardless of the nature of the solvent, but the exact degree of inhibition by the solvent can be related to the solubility of the polymer chains of lower molecular weight in the solvent, as well as to the presence of solvent ions or active solvent radicals.

Higher activation-energy requirements[28,44,69,70,87,116] are reported both for bulk polymerization and solution polymerization compared with the activation energy of the same polymer in emulsion form; values of 23,000-28,000 cal/mole have been reported for bulk and solution thermal polymerization of styrene. The data on the activation of unsaturated monomer molecules in emulsion polymerization show greater disagreement, presumably from the variations in the conditions employed by each investigation in the development of the emulsion micelles. The activation-energy values reported for thermal polymerization of styrene in emulsion have varied from 16,900-23,000 cal/mole. However, the majority of investigators favored a value between 17,500 and 22,000.

Emulsion and Suspension Polymerization.

The most common types of heterogeneous polyreaction are polymerizations in suspension and emulsion.[32,37,105] Emulsion polymerization has been discussed extensively, whereas the technical literature on suspension polymerization is largely qualitative only, except for the wide patent coverage and disclosures.[15,16,106]

The principle of suspension polymerization, or *bead* or *pearl* polymerization as it is sometimes called, is that the liquid monomer is dispersed in a nonsolvent by mechanical agitation, and polymerization then takes place inside the small suspended globules. The polymer is purer than that obtained in the presence of surface-active agents and except for a suspension stabilizer, no other ingredients are added unless the polymerization requires a catalyst or modifier. The profound influence of mechanical agitation and solvent on polymer-chain termination and the danger of coalescence of the monomer globules make this method somewhat undependable in pure scientific studies. From a commercial point of view, bead polymerization gives a purer grade of polymer and more uniform average molecular weight and is a process easily manageable. These factors combine to make bead polymerization attractive to polymer manufacturers.

In the laboratory, the polymerization is usually carried out in a three-neck flask equipped with a mercury seal stirrer, a condenser, and a thermometer for monomers with boiling points above the reaction temperature. For the polymerization of lower-boiling or gaseous monomers, electrically heated autoclaves on a shaking machine are necessary.

Two principal types of catalyst are used for this method of polymerization: hydrocarbon-soluble organic peroxides, and water-soluble inorganic peroxides or salts of peracids. The initiation temperature of styrene polymerization in dispersion, with benzoyl peroxide as a catalyst, lies above 50°C.[5,22,30,89]

The first step in suspension polymerization is an induction period during which the monomer remains liquid and no tendency to coalesce exists as long as mechanical agitation is continued. During the elementary chain growth, the polymer that forms within the dispersed globule dissolves in the remaining monomer. Plasticization of the polymer through solution renders the suspended particles very sticky and unless the agitation is sufficiently vigorous coalescence occurs through adhesion of two or more particles. The merger of sticky particles can be minimized by keeping the polymerization temperature below the softening point of the polymer and by uniformly vigorous stirring. As polymerization continues, the average molecular weight of the polymer increases and less monomer remains within the dispersed globule. The higher the average molecular weight of the polymer, the higher its softening point and, therefore, the less the tendency toward coalescence.

Disagreement over exact activation energies required for the polymerization of styrene in bulk in organic solvents and in aqueous suspension does exist, but there is general agreement that these modes of polymerization have activation energies of the same order of magnitude.[34,41,68,69,88,89] Chemists agree that a substantial decrease in activation energy requirements exists when polymerization of the monomer is carried out in emulsion.

The polymerization of styrene in aqueous dispersion is an example of bead or pearl polymerization. Note that the monomer is never completely insoluble in the suspension medium and in some cases the solubility is very high, e.g., in the bead polymerization of vinyl acetate. Therefore, in this method of carrying out a polyreaction, solution polymerization and suspension polymerization may occur simultaneously, each contributing to the complexity of the reaction.

Polymerization of vinyl compounds, either catalytically or by thermal activation, proceeds through a chain process.[33,74,90] The active centers in both types of promoted reactions are considered to be free radicals and presumably the free radical in thermal polymerization results from the collision of two monomer molecules:

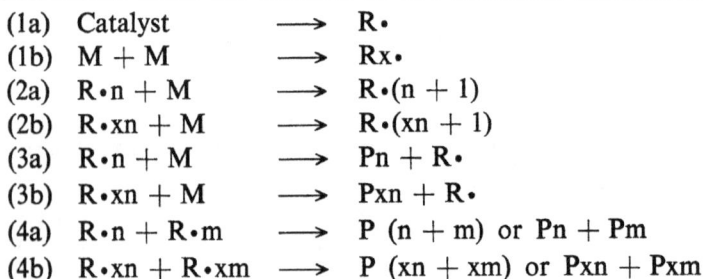

(1a)   Catalyst $\longrightarrow$ $R\cdot$
(1b) $M + M$ $\longrightarrow$ $Rx\cdot$
(2a) $R\cdot n + M$ $\longrightarrow$ $R\cdot(n + 1)$
(2b) $R\cdot xn + M$ $\longrightarrow$ $R\cdot(xn + 1)$
(3a) $R\cdot n + M$ $\longrightarrow$ $Pn + R\cdot$
(3b) $R\cdot xn + M$ $\longrightarrow$ $Pxn + R\cdot$
(4a) $R\cdot n + R\cdot m$ $\longrightarrow$ $P(n + m)$ or $Pn + Pm$
(4b) $R\cdot xn + R\cdot xm$ $\longrightarrow$ $P(xn + xm)$ or $Pxn + Pxm$

where

R· is a free radical from the decomposition of the catalyst

M is an unactivated monomer

R · x is a free radical consisting of two activated monomer molecules, which yield an ionic dimer

Pn and Pxn are polymer terminated chains

Pn + m and Pxn + xm are polymer terminated chains

R • m or R • xm are free radicals from the growing polymer chain, solvent radical, or catalyst radical

The polymerization of vinyl derivatives, whether through homogenous bulk and solution polymerization or through heterogeneous suspension and emulsion polymerization, proceeds through some scheme of free-radical chain mechanism. The previously listed elemental steps[7,29,62,74,94,98] are referred to as *activation* (1a, 1b), *propagation* (2a, 2b), *chain transfer* (3a, 3b), and *termination* (4a, 4b).

Polymerization in suspension yields a polymer with a narrower distribution of molecular weight than is likely to be found in normal bulk polymerization. A similar polymer could be produced by emulsion polymerization, but other conditions might demand a lower rate of polymerization, a lower degree of polymerization, or a relatively lower average molecular weight; these properties are not likely to be met in polymers derived by emulsion technique.

Emulsion Polymerization

The earliest study of the polymerization of olefins and diolefins dates back to 1839, when Simon transformed a styrene into a gelatinous mass under the influence of "air and heat." However, all early investigations refer entirely to the pure liquid phase, and the first ideas of using a finely divided monomer in suspension were conceived by Hofmann and Delbrück,[54] Gottlob,[42] and Berthelot.[10] The work was essentially different from what is known today as emulsion polymerization, resembling more nearly the technique of suspension polymerization.

Practical and scientific considerations have made emulsion polymerization the most valuable technique in a study of resin formation. High-percentage conversion of the monomers within a fraction of the time required by other methods, high degree of polymerization, polymer of high average molecular weight, high purity of polymer produced, improved color, a wider selection of initiators (including water soluble types), economy of operation, and many other commercial and fundamental factors are the reasons for the extensive experimentation carried on in this field during the past years.

Emulsion polymerization formulas are developed for a particular monomer or comonomer, but in general they have the composition indicated for a butadiene–styrene system [179] in Table 2—2.

TABLE 2-2

Typical Emulsion Polymerization Formulation

	Parts by Weight
Butadiene	75.0
Styrene	25.0
Water	180.0
Soap	5.0
Modifier	0.5
Catalyst	0.3

The polymer solids obtained from typical commercial latices have been found to vary between 25 and 55%. The stability of the emulsion, the particle growth, and the speed of conversion to polymer limit the production of latices containing high-polymer solids. By using synthetic surface-active agents instead of soaps, e.g., the sodium salt of an aryl–alkyl polyether sulfonate, latices as high as 60% in polymer solids have been reported.[180] Fatty-acid and rosin-acid soaps have been used in large quantities in the preparation of GR-S latices and other elastomers, but recent work on experimental latices has emphasized the use of synthetic detergents to achieve better stability against electrolyte concentrations, better storage stability, smaller particle size, and improved color.

Studies on the copolymerization of butadiene and styrene and on the polymerization of dienes in aqueous emulsion have shown that more useful elastomers are obtained when the polymerization reaction occurs in the presence of a substance known as a "modifier." Tertiary mercaptans (thiols) such as tert-hexadecyl mercaptan have been found to exert a uniform and controllable degree of modification throughout the polymerization of GR-S formula.[176] The theory of modifier action in diene polymerization assumes that the thiols can act as transfer agents for growing free-radical chains. The total effect of this behavior is to produce a more linear polymer with less branching and one that can be milled and processed like natural rubber. A reduction in the molecular weight of the polymer, together with the linear-chain character, gives an elastomer more soluble in aromatic solvents and ketones.[181,182]

Recent disclosures on fast, low-temperature formulations for rubber copolymers, resulting from investigation of the problem of initiation of the polymerization reaction, not only add to the economy of operation but, more significantly, result in consistent improvement of properties.[175] Low operating temperatures produce polymers with a lower tendency toward the formation of cross-linked polymers, higher tensile strengths, improved elongations and elastic behavior in the case of rubber copolymers and improved flex and tear resistances. Most emulsion-polymerization investigators now agree that a major improvement in the quality of an emulsion polymer can be obtained by alterations in the reaction temperature. The accomplishment of low-temperature reactions has been made possible by the intensive studies on initiation catalysts and modifiers.[176,177,178]

The physical properties of a butadiene–styrene comonomer emulsion, initiated by the diazo-thio-ether, 1-(2,4-dimethylbenzenediazo-mercapto) naphthalene and activated by potassium ferricyanide at $-18°C$, yielded a copolymer with physical properties considerably superior to those exhibited by GR-S at 50°C.[175] No theory on the mechanism of diazothio-ether initiation has been developed, so that an explanation of this efficiency cannot be offered.

Table 2-3 summarizes a report on emulsion polymerization formulas for synthetic rubber, including low-temperature polymerizations in redox systems. The operation of oxidation–reduction catalysis differs from the classical methods that employ persulfate salts and peroxides initiation chiefly in the temperature of polymerization. To carry out low-temperature conversion, freezing-point depressants such as methanol, glycol, and glycerol are used. The depressants retard the polymerization reaction, but the process remains quite practicable even in plant-size operations.

The development of emulsion formulas has been associated with some empiricism and part of the difficulty probably stems from the controversy over the loci of the reaction in emulsion systems.[36,46,69,72,73,87,96,112]

W.D. Harkins[46] has advanced a general theory of the mechanism of emulsion polymerization that differs from all previous work. Up to the time of Harkins' theory, it was held questionable that emulsion polymerization was initiated in the aqueous phase[36,37,38,112] and not at the interface or in the monomer phase. The older theory, which assumed that the monomer droplet itself polymerizes, is disregarded by all recent investigators. The principal difficulty seems to arise from a definition of the nature of the aqueous phase. Generally, writers before Harkins appear to assume the position of Fikentscher,[32] regarding the aqueous phase with minor modification related to reaction mechanism. By considering (with Fikentscher) the aqueous phase to include oil-emulsion droplets, polymer–latex particles, and soap micelles, the investigators fail to define the true aqueous phase and

TABLE 2-3

Emulsion Polymerization Formulas *†

	1	2	3	4	5	6	7	8	9
Butadiene	71.5	70	70	55	72	70	70	70	70
Styrene	28.5	30	30	—	28	30	30	30	30
Acrylonitrile	—	—	—	45	—	—	—	—	—
Total Distilled Water	180	250	175	250	180	180	126	126	192
C.P. Methanol	—	—	75	—	—	—	—	—	48
Glycerol	—	—	—	—	—	—	54	54	—
Sodium Myristate	—	—	—	5	—	—	—	—	—
Rubber Reserve Soap	—	—	—	—	1.2	—	—	—	—
Dresinate [a] No. 731 (Dry Basis)	4.68	5	—	—	3.5	—	—	—	—
Potassium Oleate	—	—	5	—	—	5	5	5	—
Potassium Laurate	—	—	—	—	—	—	—	—	5
Soap Solution, pH	10	10.3	10	—	10	10	10	10	—
Soap Solution, % Neutralized	—	—	—	85	—	—	—	—	95
Potassium Hydroxide (Excess)	—	—	0.1	—	—	—	0.1	—	—
Bi-Isopropylxanthogen	—	—	—	0.3	—	—	—	—	—
n-Dodecyl Mercaptan (DDM)	0.5	—	—	—	—	—	—	—	—
Mixed tert-Mercaptans [b]	—	0.4	—	—	0.4	0.25	0.17	0.25	0.25
n-Hexadecyl Mercaptan	—	—	1	—	—	—	—	—	—

	1	2	3	4	5	6	7	8	9
Hydrogen Peroxide (100%)	0.3	—	—	0.35	—	—	—	—	—
Potassium Persulfate	—	—	—	—	0.15	0.076	0.15	0.34	0.167
Cumene Hydroperoxide (100%)	—	9.3	—	—	—	—	—	—	—
(MDN)[c]	—	—	—	—	—	—	—	—	—
(a-XDN)[d]	—	—	0.3	—	—	—	—	—	—
Potassium Ferricyanide	—	0.3	0.3	—	—	—	—	—	—
Sodium Pyrophosphate $10H_2O$	—	—	—	0.65	1	1	1	1	0.446
Ferrous Sulfate $7H_2O$	—	—	—	—	0.1	0.111	0.1	0.44	0.278
Ferric Pyrophosphate (Solution)	—	—	—	0.1	—	—	—	—	—
Glucose	—	—	—	—	1	—	—	—	—
Levulose	—	—	—	—	—	1	—	—	—
Sorbose	—	—	—	—	—	—	3	—	—
Trisodium Phosphate $12H_2O$	—	—	0.5	—	—	—	—	—	—
Potassium Chloride	—	—	—	—	—	—	—	—	(0.5)
Temperature, °C	50	5	-18	30	30	15	-10	-10	-10
Time, Hours	14	3.5	31	23	4	7	46	5.1	7.6
Conversion, %	72	60	58	99	72	61	73	60	43

* All quantities are given as parts by weight unless otherwise specified in first column.

† Ind. Eng. Chem., **41**, 987 (1949).

a Sodium salt (soap) of disproportionated rosin acid.

b tert—$C_{12}SH$, 60 parts; tert—$C_{14}SH$, 20 parts; and tert—$C_{16}SH$, 20 parts, abbreviated MTM-4.

c 2-(4-Methoxybenzenediazomercapto) naphthalene.

d 1-(2,4-Dimethylbenzenediazomercapto) naphthalene.

do not specify the locus of the reaction.

According to Harkins, the principal function of the monomer emulsion droplets is to act as a storehouse of monomer from which its molecules diffuse into the aqueous phase, and from this, into either soap micelles or polymer–monomer latex particles. The aqueous phase is considered limited only to the aqueous true solution, exclusive of oil-emulsion droplets, polymer–latex particles, and soap micelles.

The two principal loci of polymerization[46] and two minor loci whose polymerization effect depends on the effectiveness of the principal follow:

(a) *Principal locus for the initiation of polymer particle nuclei:* Solubilized monomer is completely surrounded by a monolayer of soap molecules (about 15-25Å).

(b) *Principal locus for formation of polymer:* Nearly all of the polymer-particle nuclei are formed by polymerization of the monomer within the soap micelles from (a). As the polymerization proceeds, the polymer molecules grow and appear in the true aqueous phase as small polymer-particle nuclei, eventually exceeding the physical dimensions of the soap micelle. The activated polymer nuclei combine with any monomer molecules that approach their effective kinetic field and thus propagation of the chain continues. The activated nuclei can be terminated, as discussed previously, by collision with a solvent radical or atom, with a catalyst radical, or with other activated nuclei.

(c) *Minor locus for formation of polymer particle nuclei:* Any dissolved monomer present in the true aqueous phase may be activated by the presence of soluble catalysts and possibly promoted through other ions—organic or inorganic—such as mercaptans and thiosulfates. This locus becomes an important contributor to polymerization in the absence of any soap.

(d) *Minor locus for the initiation of polymer particle nuclei:* The comparatively larger emulsion droplets can behave as a locus for the initiation of polymer nuclei at a very low rate. At first, the conditions favor diffusion through the soap area around the emulsion droplet to the outer aqueous phase, but as a consequence of polymerization, soap disappears from the aqueous phase and later becomes adsorbed around the emulsion droplets and other polymer particles. The whole effect causes diminishing reactivity because a reduction of the total area of the surface of the monomer emulsion droplets allows fewer and fewer molecules of monomer to diffuse into the aqueous phase. The inhibition of monomer diffusion favors the formation of activated nuclei within the emulsion droplet.

Fryling and Harrington[37,126] reported that undispersed acrylonitrile in a water system and in the presence of a watersoluble accelerator substantiates the view that polymerization is initiated in the aqueous phase. They find additional support in the observation that a volatile monomer polymerizes even if it is not in direct contact with the aqueous phase. Their conclusions indicate that in emulsion polymerization the soap assists the solubilization of the monomers and that responsibility for solubilization may be interpreted in terms of the micellar structure of the soap in accordance with the expressed views of McBain and others.[51,52,53,75] Apparently, a proportional numerical relationship exists between polymerization velocity and soap concentrations[37]—a relationship accounted for not by the theory that polymerization occurs at the interface, but rather by the improved solubilization of the monomer. The confusion, according to these investigations, results from subsequent polymer growth, which presumably occurs in the organic phase. Consequently, a growing polymer chain must penetrate the phase interface in order to grow to relatively high molecular weights, but it is likely to be terminated in the course of its random movement and it is possible for the termination reaction to occur near or at the interface.

Harkins considers each double layer of soap as an independent micelle, capable of free movement and uninhibited by a pile of double layers as postulated in earlier theories. The increase in viscosity at high concentrations of soap is apparently due to a growth of the micelle and not to a further development of lamellar structure. Under either circumstance, the soap micelle plays an important role in emulsion polymerization, because it dissolves the monomer. Polymerization begins while the monomer is inside the soap micelle, and the quantity of micellar soap determines the number of polymer particles formed and thus the rate of polymerization.

Solubilization of the monomer appears to be intimately associated with the hydrolysis of the soap.[37] The pH of an aqueous soap solution decreases on solution of monomer and increases again as the fatty acid resulting from hydrolysis is removed from the aqueous phase by solution in the organic monomer phase.

The hydrolysis equilibrium of soap may be represented by

$$RCOO^- + Na^+ + H_2O \rightleftharpoons RCOOH + Na^+ + OH^-$$

McBain suggested that organic micelle formation would be accompanied by a decrease in hydrolysis, and accordingly Fryling proposed the following equilibrium as an explanation of the competing reaction:

$$RCOO^- + Na^+ + nM \rightleftharpoons [(M)n \cdot RCOONa]$$

where M is any relatively water-insoluble monomer.

Equilibrium does occur to a considerable extent and the shift continues to the right after solubilization, as a result of solution of the fatty acid in the organic phase. An increase in pH occurs invariably on further addition of the organic monomer. A decrease in pH[37,87] in the course of emulsion polymerization is also reported. This effect may be due to degradation products of the polymer, to oxidation products from the peroxide, or to the formation of acid salts from ionic promoters. Although these salts contribute to a lowering of the pH, many now believe that the process is more complex, involving a regeneration of fatty acid as the monomer passes into the polymer.

All these facts make it probable that in heterogeneous emulsion systems the formation of active nuclei occurs chiefly in that part of the monomer solubilized in the soap micelle, to a lesser degree in the bulk phase of the monomer, and to a questionable degree in the strictly water phase. The rate of polymerization is affected by temperature and the concentration of the soap, aside from any consideration of catalyst and promoter. In general an increase in soap concentration increases the polymerization rate. The activation effect of soap on styrene polymerization has been confirmed,[36,87,112] and in the presence of potassium persulfate the activation energy for styrene polymerization was found to be about 17,000 cal/mole, or about 8,000 cal less than that required for polymerization in bulk or homogeneous solution.

Previous investigations[34,87] have shown that the length of the inhibition period is dependent on the temperature. Some investigators [34,44] have stated that the energy of activation for a polymerizable vinyl compound can be determined from the length of the induction period measured at different temperatures, whereas others[87,126] divide the induction period into a subperiod and an inhibition period, and consider that the growth of active nuclei to a significant length begins immediately after the inhibition period. These investigators measure the energy of activation by the length of the that period—the time necessary for the monomer to free itself from the influence of any inhibitor so that the formation of a long chain can begin.

Active centers are formed during the inhibition period, but are destroyed by the inhibitor almost immediately. As yet, no investigator has completely freed his experiments from the effects of inhibitors, which may be traces of stabilizers or oxidation products from the catalyst, or may arise from collision with nearby activated and unactivated monomer molecules. Following the inhibition period, therefore, the formation of the polymer can be more accurately measured. Active centers form in large numbers in the soap micelles at this time and propagation starts in the solubilized monomer, as well as in the polymer particles swollen in the monomer.

According to Kolthoff and Dale,[65] the rate of conversion of styrene into polystyrene via emulsion polymerization, excepting the inhibition period, is proportional to the square root of the weight of potassium persulfate present as the catalyst in the system. In this respect, the kinetics of the emulsion polymerization of styrene are comparable to that of the bulk and solution polymerization.

Assuming that the catalyst dissociates into free radicals which in turn react with the monomer to initiate a chain, then the mechanism can be illustrated as follows[65]:

(1) $A_2 \longrightarrow 2A \cdot$ (Dissociation of catalyst)
(2) $A \cdot + M \longrightarrow AM \cdot$ (Active nucleus)
(3) $AM \cdot + nM \longrightarrow AM \cdot (n + 1)$ (Propagation)
(4) $P \cdot n + P \cdot m \longrightarrow P_n P_m$ (Termination)

Many intricate and sometimes conflicting statements on emulsion polymerization are made; nevertheless, some generally accepted deductions can be stated:

Emulsion polymerization proceeds more rapidly than homogeneous bulk or solution polymerization;

Organic monomers are solubilized by soap and thus, one may expect that the concentration of the monomer in the soap solution will influence the reaction rate;

The average molecular weight of the polymers formed by emulsion technique is higher than that of polymers resulting from homogeneous polymerization.

Many more controversial statements can be found in the literature despite the attempt of investigators[2,36,37,46,50,112,130,131] to clarify the subject. The nature of an aqueous solution of the colloidal soap electrolyte is very complicated in itself. Whether polymerization takes place in the true aqueous phase, soap micelle, or emulsion droplets has been the subject of many papers.[32,36,37,46,47,48,49,50]

Figure 2-1 is a typical time-conversion study on the emulsion polymerization of styrene, illustrating a slow inhibition period, a rapidly increasing propagation period, the steady state, and finally the prominent termination reactivity.

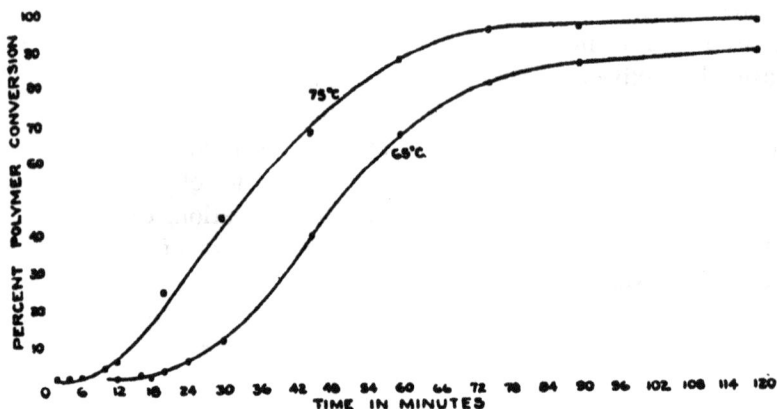

Fig. 2-1. Time-conversion curves for styrene polymerized in emulsion.

MOLECULAR WEIGHTS

At the present time, determination of the molecular weight of a high polymer from viscosity measurements alone is impossible because each solute–solvent system has its own characteristics with respect to molecular shape, molecular volume, and solvent behavior. Macromolecules, when sufficiently branched, can be round; when linear, their shape can vary from rigid rods to randomly-kinked and coiled chains.

A simple relationship between viscosity and molecular weight, which would be applicable to new solute-solvent systems, without the necessity of evaluating the molecular weight from the so-called absolute methods of osmotic pressure and the ultra-centrifuge, is still being sought. Cryoscopic methods for the determination of molecular weight much greater than 5,000 are inaccurate, [135] because the change in freezing point, for example, is too small to be measured with sufficient precision at low concentrations of polymer solution.

Staudinger[98] has tried unsuccessfully to show some constant relation of viscosity to concentration and molecular weight. The results were inconclusive, because variations in Staudinger's K_m value ranged from

$(7.4—1.7)10^{-4}$ in the polymeric range of average molecular weights from
500—13,000. Viscosity and molecular-weight relationships for higher
polymers gave Staudinger[99] further deviations for K_m from $(1.2—0.4)10^{-4}$. The unsatisfactory data are presently attributed to the false
assumption by Staudinger that the ratio of specific viscosity to con-
centration is constant for all large molecules in a homologous series,
and that η_{sp}/C is proportional to the number of rod-like and randomly-kinked solute molecules. Having determined the molecular weight
by the usual cryoscopic method, Staudinger found a value of K_m that
was not valid throughout a heterogeneous polymer. The well-known
Staudinger equation is

$$M = \frac{\eta_{sp}}{CK_m}$$

where M is the molecular weight of the solute, K_m is the proportionality constant, η_{sp} is the specific viscosity, and C is the concentration
expressed in base moles per liter of solution, as grams per 100 cc of
solution, or as grams per liter.

The literature on the molecular weight is not entirely satisfactory
because the various methods acceptable for this determination do not
give identical average values, but the results are rather affected by
the shape of the molecules and the type and concentration of the
solutions used in the experiment.

The viscometric method lends itself to simple measurements better
than does any other procedure. The viscosity of a liquid is a measure
of its resistance to flow, which has been called its internal friction.
The coefficient of viscosity is defined as the shearing force per unit
area required to maintain a unit velocity gradient between two parallel
plates at a constant distance. The equation can be expressed as
follows:

$$\eta = \sigma/(du/dy)$$

where η is the coefficient of viscosity of the liquid, σ is the shearing
force, u is the velocity in the direction of flow, and y is the distance
in the direction normal to the planes of shear in the liquid.

In most simple liquids, where streamline flow exists without turbulence, the velocity gradient is a linear function of the shearing force;
ideal liquids (which show this linear relationship) are in general known
as Newtonian liquids. The most accurate method of measuring the
viscosity of a Newtonian liquid is based on the Hagen–Poiseuille
equation, which expresses the principal law of flow:

$$\eta = \frac{\pi r^4 P t}{8 V l}$$

where V is the volume of liquid flowing in time t, P is the pressure, r is the radius of the capillary tube, and l is the length of the tube.

However, the study of polymeric substances is concerned with non-Newtonian behavior. The two departures from the normal behavior are associated with solutions and sols of high polymeric substances. A normal liquid is characterized by a constant relation between η and shear stress. Elastic solutions have a larger η at small shear stress, and η approaches the Newtonian behavior with increasing shear stress. Superfluid solutions show normal behavior under the smallest shear stresses, but with increasing stress, they depart more and more from the Newtonian behavior. Both elastic and superfluid effects are found in polymeric solutions. Concentrated solutions of substances of high molecular weight may behave as elastic solutions —a phenomenon called thixotropy—without reference to the possible existence of solute–solvent arrangement as an explanation. Superfluidity has been noted in dilute solutions of kinked molecules that become stretched or flattened by flow.

Einstein[148] pointed out that if the suspensions are not very concentrated, the specific viscosity is independent of the size of the particles and depends only on the volume fraction of rigid spheres, according to the equation:

$$\eta_{sp} = 2.5 V_2$$

where V_2 is the volume fraction of the solute. This classical equation is based on assumption of ideal conditions, where there is no disturbance of the particles either by solvent or other solute particles.

Guth, Gold, and Simha[148] extended Einstein's hydrodynamic treatment to examine the dependence of internal friction on concentration at higher concentrations of large spherical solute particles. For higher concentrations, Einstein's relationship is no longer linear, but may be expressed as a power series as follows:

$$\eta_{sp} = 2.5 V_2 + 14.1 V^2_2 + \dots\dots$$

Other equations have been derived for ellipsoidal[149] particles, for rigid rodlike pieces and disks,[150] and randomly-kinked, linear solute molecules.[151] With elongated rods, excessive viscosities were always found, even at the highest measurable dilutions, indicating some form of interaction beyond the validity of formulas. In highly dilute solutions of high-polymeric substances with viscosity behavior approaching ideal conditions, the determinations are complicated by isolated flexible chains in a curled or tangled state, and simultaneously under the influence of Brownian movement.

From various observations in a field still in the development stage, it has been shown that the ratio of specific viscosity to concentration —η_{sp}/C—for a homologous series is (a) constant for spherical solute molecules (Einstein's law); (b) proportional to n^2 for rodlike solute molecules; and (c) proportional to n for randomly-kinked solute molecules, where n is the number of solute molecules (Staudinger relationship[98]) per unit of solution.

More recently, the dependence of viscosity on the swelling of the solute particle has been studied. Kuhn, Alfrey, and Gee[152] report that a fair solvent might swell the polymer particle to such an extent that the interpretation of data would be meaningless unless a swelling factor[153] is introduced.

When a substance of high molecular weight is dissolved in a solvent of low molecular weight, the viscosity of the solution η is greater than that of the pure solvent, η_0. The increase in viscosity value is called the relative viscosity η_r, from which the specific viscosity η_{sp} is obtained:

$$\eta_r = \frac{\eta}{\eta_0} = 1 + aC + a_2C^2 + a_3C^3 + \ldots$$
$$\eta_{sp} = \eta_r - 1 = aC + a_2C^2 + a_3C^3 + \ldots$$

The increase per unit concentration has been termed the *reduced* viscosity:

$$\frac{\eta_{sp}}{C} = a + a_2C + a_3C^2 + \ldots$$

When working with dilute solutions under 1%, one can neglect the higher terms of these equations. The equation for the reduced viscosity is usually given as

$$\frac{\eta_{sp}}{C} = a + bC$$

where $b = a_2$.

The reduced viscosity is the quantity that Staudinger[98] used originally as a means of estimating molecular weights:

$$\frac{\eta_{sp}}{C} = K_m M$$

Investigators now agree that this equation suggests a proportionality that does not exist. The equation does not allow for any molecular peculiarities or solvent–solute activity.

Huggins[151] and Kraemer[154] evaluated the constants a and b in the equation for the reduced viscosity and found them to be

$$a=\left(\frac{\eta_{sp}}{C}\right)_{C=0} \text{ and } b=k'[\eta]^2$$

where $[\eta]$ is the intrinsic viscosity $\left(\frac{\eta_{sp}}{C}\right)_{C=0}$.

The equation for reduced viscosity became

$$\frac{\eta_{sp}}{C}=\left(\frac{\eta_{sp}}{C}\right)_{C=0}+k'\left(\frac{\eta_{sp}}{C}\right)^2_{C=0}C$$

Kraemer called the limiting value of (specific viscosity)/(concentration at infinite dilution) the intrinsic viscosity. His method is today the accepted procedure for relating solute concentration to viscosity and viscosity to molecular weight via the Mark equation.

Kraemer's equation for intrinsic viscosity is usually given as follows:

$$\frac{\eta_{sp}}{C}=[\eta]+k'[\eta]^2C$$

where

$$[\eta]=\left(\frac{\eta_{sp}}{C}\right)_{C=0}$$

The value k' was suggested by Huggins [151] as a constant for a given solute-solvent system dependent on the size, shape, and cohesional properties of both solvent molecules and solute molecules, but reportedly not on the length of the solute-molecule chain. Accordingly, once the correct value for k' is known, a single viscosity determination suffices for the calculation of intrinsic viscosity. Graphically, the value of k' is evaluated from the slope of the line obtained when the reduced viscosity is extrapolated to zero concentration:

$$k'[\eta]^2=\text{slope}$$

Huggins was supported by Martin [155] on the constancy of k' for a given solute–solvent system. However, several investigators[156,157,158] have shown that k' is dependent on molecular-weight distribution in the same solvent, and thus it is not constant unless a fractionated polymer is used in the determination.

The value of k' for many linear polymers in good solvents [135] is about 0.38, but it may be different if the polymer is branched or cyclized, if the solvent is poor, or if there is chemical interaction between solvent and solute. Apparently, k' increases with increasing width of molecular-weight distribution. As most values of k' for fractionated high polymers in good solvents vary from 0.35 to 0.41,[135] normal polymerization distribution of molecular weights appears to exist within this range. A high value of k' should indicate a poorly fractionated polymer, a polymer with a wide distribution of molecular weights, or possibly a poor solvent; a low value for k' signifies a predominance of short-chain molecules, but not necessarily a poor molecular-weight distribution.

The only practical way of utilizing k' in accordance with Kraemer's practice of calculating intrinsic viscosities for a given solute-solvent system would be to determine k' from a group of fractionated samples and average the values for a general viscosity–molecular-weight average distribution. The value of k' obtained from unfractionated polymers would be unreliable, unless many intrinsic viscosities were graphically obtained from a good distribution of samples during various stages of polymerization.

The Staudinger Equation

Staudinger[98] suggested that the reduced viscosity η_{sp}/C was proportional to the molecular weight:

$$\frac{\eta_{sp}}{C} = K_m M$$

He recognized the necessity of evaluating viscosity in those regions where solute association would not be disturbing, and proposed the determination of the viscosity at a "sufficiently low" concentration. Staudinger worked originally in the low region of molecular weights of polystyrene and it is understandable how he might have concluded that the reduced viscosity was proportional only to the number of solute molecules through a proportionality constant K_m, computed on samples from cryoscopic or osmotic-pressure measurements.

Staudinger assumed that his viscosity law held strictly for linear polymers and that any deviations in K_m were due to branching of the polymers. Actually, his assumption has poor scientific foundation; branching does alter the viscosity of a polymer with a molecular weight similar to that of a linear polymer. Meyer[159] showed that the value of K_m decreased even though no branching in a polymer was evident. Staudinger's equation holds approximately for certain dilute systems of linear polymers, but it is not of general value over a wide range of molecular weights. Many now believe that deviations [160] from the law result from variation in the compactness of kinking of unbranched-chain molecules. Simha's work [150] shows that the law is invalid for systems containing stiff rodshaped molecules, for which it was stated that the reduced viscosity was proportional to the square of their number.

The Mark Equation

Mark[161] proposed a nonlinear relationship between intrinsic viscosity and molecular weight, and his equation has been used by other investigators[128,129,133,140,162,163] in the study of old and new polymers, particularly polystyrene. Mark's equation is expressed as

$$[\eta] = KM^a$$

Houwink[133] found the exponent a to be as low as 0.5 for randomly-kinked molecules, but generally the exponent approximated 1.0 for such molecules, as proposed by Staudinger. In accordance with the theory, a is greater than 1.0 for rigid rodlike molecules[140]. The exponent a, therefore, depends on the polymer and solvent structure.

If the average value of k' is known, a single viscosity measurement is sufficient to determine $[\eta]$ for the particular system, from which, with known values of K and a, the average viscosity molecular weight can be calculated.

Like k', the value of the Mark equation is questionable when applied to a heterogeneous system of unfractionated polymers because the significance of a is lost. Nevertheless, the practice of utilizing the equation empirically to determine the magnitude of the viscosity average molecular weight is continued in scientific studies, and in industry as a means of product control.

The constants K and a are determined from the logarithmic form of the Mark equation

$$\log [\eta] = \log K + a \log M$$

which is the equation of a straight line; $\log K$ is the intercept and a is the slope. To plot the line, absolute values of M are determined by a number-average molecular-weight method such as osmotic pressure[147] or freezing-point depression for low molecular weight polymers,[164,165] end-group determinations,[166] ultracentrifugal analysis,[167] and more recently[168] by means of the intensity of scattered light when a beam of unpolarized light passes through the solution of the high polymer.

The values for intrinsic viscosity are generally determined for the same polymer fractions used for molecular-weight determination by extrapolating the values of the reduced viscosity at several concentrations to zero concentration.

Log $[\eta]$ is next plotted against log M and the intercept K and slope *a* are found from the graph. Although the values of K and a are said to be constant for a given solute–solvent system, agreement between investigators is still wanting and one can report only the more acceptable values. The constants may depend somewhat on the method by which the polymer is prepared.[140]

Under the most suitable conditions, unfortunately, the results of molecular-weight determinations remain average values, because high polymers are not pure substances but are mixtures of hundreds of pure substances that are members of the same homologous series. Investigators in the field of high-polymer measurements have sought to minimize the discrepancy, or at least to render the polymer more homogeneous through fractional separation of molecular chains by means of solubility differences.

The measurement of viscosity has been discussed at length, because it is a simple means of quality control for industrial consumers. In fact, some manufacturers of vinyl resins offer intrinsic viscosity values along with a specified grade. This procedure may lead to practical data correlating physical properties and viscometric values. Certainly, maximum film properties are associated with the resin of highest average molecular weight, i.e., the higher molecular-weight resin of a given series exhibits better tensile strength and produces a tougher and more flexible film, improved chemical and solvent resistance, and a better modulus, than does resin of lower molecular weight.

For control purposes, the Staudinger equation is as worthy of consideration as Mark's equation, because the values to be obtained are merely relative to those of some other sample of a similar resin. The selection of solvent, temperature, concentration, and constants should be consistent with the available literature,[132,135,141,142,143,148,150, 151,154,160,164,165] Viscosity determinations are usually made in Ostwald–Fenske viscosity pipettes, and readings collected at constant temperature for solutions with concentrations of 1.0, 0.75, 0.50, 0.40, and 0.20 g per 100 cc of solution. By plotting concentration against the reduced viscosity η_{sp}/C, intrinsic viscosity data can be evaluated by extrapolating the curves to zero concentration. The values collected over a period of time are not to be used for absolute evaluation, but should be helpful in interpreting variations from lot to lot of a given resin.

For the scientific reader who may wish to carry out absolute measurements on number-average molecular weights, references are made to osmotic-height methods,[146,147] freezing-point depression,

[164,165] high-polymer end-group determinations,[166] ultracentrifugal analysis, [167] and to light-scattering methods.[168] A discussion of each of these experimental methods is beyond the scope and purpose of this chapter.

SYNTHETIC RESINS

Chemists naturally become overzealous in discussing resins and think mainly in terms of the synthetic product, but development men know that the position and importance of the natural resins or rubbers cannot be overlooked. It is true that the natural products have been permanently supplanted by synthetic resins; nevertheless, the natural substance remains a challenge even today in many of the industrial applications. The stabilization of synthetic resins and rubbers has finally become a reality, but in some industries the situation is still a compromise between quality and economy. An example is the artificial leather industry, where the contributions made by natural rubber latex have not been totally replaced, even though the use of synthetic rubber latex in the saturation of loosely-felted paper sheets has rapidly expanded.

Differences in tear strength, thermoplasticity, rate of cure, plasticization, tensile strength, modulus, and elongation between GR-S latex and natural-rubber latex films can be appreciated, if it is realized that none of the synthetic products are synthetic duplicates of the natural product, but rather are new material similar to a particular natural substance. So far, no true synthetic rubber has been made. The successes achieved merely emphasize the magnitude of possibilities open to the organic chemist. "Cold" rubber is an improvement over GR-S—because a smaller proportion of low molecular weight polymer is formed. The superior physical behavior of cold rubber is reportedly also due to the increased linearity of the polymer chains.

Synthetic resins were developed as a consequence of the search for improved materials. The natural polymers such as casein, rosin, and manila and copal gums, which still serve the paper industry in many capacities, have been known to be deficient in many chemical and physical properties. Water resistance, water repellency, acid number, color, abrasion resistance, film strength, electrical properties, and several other properties of the natural resins are considerably below the specifications of industrial consumers. The tanning of glue and casein, the hydrogenation of rosin, and the purposeful oxidation of oils were the chemist's and the skilled artisan's first attempts to modify nature's products, and the continued success of these processes is certainly evidence of man's accomplishments.

The chemist has not been able to duplicate fully nature's offerings, but has emulated its patient methods by the science of polymerization. The new approach permitted modification of the properties of resins practically at will. The changing of polymerization time, temperature, or catalyst; the addition of a second monomer for copolymerization; the introduction of substituent groups on the monomer; and other techniques are now in the hands of the preparative chemist.

Fig. 2-2.   Polymerization kettle (*Courtesy of Pfaudler Co.*).

Review of the physical properties of synthetic resins in terms of their chemical structure [169] and broad division of the various types of resin into three classes is customary: (1) polymers showing crystal structure and strong intermolecular forces, such as vinyon and nylon; (2) polymers having a randomly coiled arrangement at rest and assuming a definite intramolecular alignment when oriented, such as Buna S, Buna N, polychloroprene, and other elastomers; and (3) polymers having properties somewhere in between the first two types, showing crystal structure when cured and intermediate intermolecular forces.

The methods of organic synthesis make it possible to alter intermolecular and intramolecular forces within the polymer by placing substituent groups on the monomer that either cause a stiffening of the residual polymer or yield a polymer with considerable rubberiness. The technique has been called "internal plasticization" in contrast with mechanical plasticization by the addition of chemical plasticizers to the polymerized resin. Internal plasticization by copolymerization has been successful in the production of synthetic rubberlike polymers, starting with an otherwise rigid, polymer-producing styrene monomer.[170,171] Nevertheless, the effect of substituent groups on flexibility and stiffness appears to be more fundamental than the utilization of a second monomer in a polymerization procedure.

Any factor that causes a weakening in intermolecular forces—for example, the introduction of hydrocarbon apolar groups, the presence of a heterochain in the main polymer, or the introduction of a comonomer—generally results in increased flexibility because the regularity of the principal polymer chain and the pure crystallinity are disrupted, and dipole interaction can be diminished appreciably, if present at all. Freedom from strong inter- or intramolecular bonds generally results in improved flexibility. Unsaturation on the main carbon chain should improve the relative flexbility of a polymer, especially if *cis* and *trans* isomers are possible.

The forces that oppose elasticity and flexibility should cause stiffness and embrittlement. Polyethylene, for example, is somewhat flexible, but the introduction of a polarizable phenyl group increases intermolecular attraction and produces steric hindrance. Polystyrene, therefore, is brittle and stiff. The presence of polar groups, such as $-OH$, $-Cl$, $-Br$, $-COOR$, and $-CN$, tends to increase electrical dipole forces, yielding resins of rigid configuration, high tensile strength, poor elongation, and high modulus.

In the evergrowing list of publications on polymerization, one factor consistently emerges concerning the relationship between the molecular weight and the softness, flexibility, and elongation—the lower the molecular weight, the less rigid are the polymeric chains. Apparently, there is less complexity of structure in the lower molecular weight resin, as well as greater possibility of slippage between the molecular chains and greater ease of disruption, an observation true of all types of resins. A lower molecular weight form of a particular resin is found to be softer, tackier, and lower in tensile strength, or more easily elongated compared with a higher molecular weight type. The strong intermolecular and intramolecular associations of any resin, and also the less mobile chains of the higher molecular weight forms of a given resin, render the resins less soluble and less compatible with other compounding substances such as plasticizers, oils, and waxes.

The rigid polymers have such valuable physical and chemical properties, however, that a great deal of development work has been carried out in industrial laboratories to evaluate the proper mode of plasticization. The problem of plasticizing tridimensional resins, such as urea–formaldehyde and phenol–formaldehyde, is a difficult one, although some successful work with phenolics and acrylonitrile–butadiene copolymers has been done.[172] Plasticization of addition polymers and copolymers was extensively investigated during World War II and a series of papers that offers a review of the literature on the plasticizing effects of various known substances has been published.[173,183]

The proper selection of a chemical plasticizer involves a combination of several factors, such as compatibility, permanency, efficiency, toxicity, color, ease of dispersion, odor, cost, and low-temperature flexibility. A special project on the saturation of paper with a high molecular weight emulsion of polyvinyl acetate, was also concerned with the permanency of the plasticizers used in flexibilizing the resin. Sheets of paper were treated with the plasticized resin emulsion and aged at 78°C in an oven equipped for air circulation. The permanency of the system was evaluated indirectly by measuring the increase in stiffness of the flexible sheet and by recording the percentage increase in stiffness with aging. The same dry weight plasticizer-resin relationship was used in the series of tests listed in Table 2—4.

The list of plasticizers in the table represents only a few of those evaluated, but it does serve to indicate the differences to be expected. However, the results are based on a system including paper fibrils; also the irregular surface of the fibers may assist in the migration and

TABLE 2-4

Percent Change in Stiffness with Different Plasticizers

	Hours			
	0-24	24-72	72-120	120-168
Di-butoxy Ethyl Phthalate	+36.5	+ 61.0	+ 65.0	+ 65.0
Triethylene Glycol Di-2- Ethylhexoate	+18.0	+ 55.0	+ 76.0	+120.0
Tributoxyethyl Phosphate	− 3.6	− 2.6	+ 48.0	+ 60.0
Butyl Phthalyl Butyl Glycolate	+13.0	+ 34.0	+115.0	+135.0
Tricresyl Phosphate	+12.5	+ 18.0	+ 94.0	+ 96.0
Bis-(Diethylene Glycol Monoethyl Ether) Phthalate	+10.0	+ 17.0	+ 46.0	+ 72.0
Tributyl Citrate	+19.0	+ 38.0	+115.0	+150.0
Methyl Cellosolve Acetyl Ricinoleate	+ 2.0	+ 8.5	+140.0	+200.0
Ethyl Phthalyl Ethyl Glycolate	+18.0	+ 65.0	+245.0	+360.0
Triethylene Glycol Di-2- Ethylbutyrate	+37.0	+110.0	+305.0	+475.0
Dibutyl Sebacate	+27.0	+ 35.0	+195.0	+250.0
Methyl Phthalyl Ethyl Glycolate	+34.0	+ 68.0	+230.0	+390.0
Dicapryl Phthalate	−17.0	− 16.0	+ 80.0	+ 81.0
Fatty Acid Amide	+ 2.0	+ 11.0	+ 85.0	+155.0
Dibutyl Phthalate	+63.0	+190.0	+580.0	+770.0
Diamyl Phthalate	+48.0	+180.0	+540.0	+740.0

removal of plasticizer from the surrounding resin. Furthermore, the results are concerned with only one factor—permanency. Several others are important, so that the selection of a suitable plasticizer becomes at best a compromise; sometimes common practice ensures multiple plasticizers in a given formula.

Development men studying the practical applications of resins generally are concerned with many specific properties or specifications not found in the tables of contents of books on resins or even in tables on plasticizer permanency. Trial-and-error research is still quite prevalent in the practical evaluation of a resin formula. Even with the necessary information from the literature, the final decision on a resin formulation still depends on results obtained from actual tests.*

---

* A substantial portion of the literature review in this chapter was made in connection with a Doctoral dissertation at Clark University, Worcester, Massachusetts.

REFERENCES

1. J. J. Berzelius, *Jahresber.*, **12**, 64 (1833).
2. J. Abere, G. Goldfinger, H. Mark, and H. Naidus, *J. Chem. Phys.* **11**, 379 (1943).
3. J. Abere, G. Goldfinger, H. Naidus, and H. Mark, *J. Phys. Chem.* **49**, 211 (1945).
3a. J. Abere, G. Goldfinger, et al., *Ann. N. Y. Acad. Sci.* **44**, 277 (1943).
4. T. Alfrey and C. C. Price, *J. Polym. Sci.* **2**, 101 (1947).
5. H. N. Algea, J. J. Gartland, and H. R. Graham, *Ind. Eng. Chem.* **34**, 458 (1942).
6. J. L. Bolland, *Proc. Roy. Soc.* (London), **A178**, 24 (1941)
7. P. D. Bartlett and S. G. Cohen, *J. Am. Chem. Soc.* **65**, 543 (1943).
8. P. D. Bartlett and R. Altschul, *ibid.*, **67**, 812, 816 (1945).
9. J. H. Baxendale, M. G. Evans and G. S. Park, *Trans. Faraday Soc.* **52**, 155 (1946).
10. M. Berthelot, *Bull. Soc. Chim.* **6**, 294 (1866).
11. A. T. Bloniquist, J. R. Johnson, and H. J. Sykes, *J. Am. Chem. Soc.* **65**, 2466 (1943).
12. J. W. Breitenbach and H. Rudorfer, *Monatsh.* **70**, 37 (1937).
13. J. W. Breitenbach and R. Raff, *Monatsh.* **69**, 1107 (1936).
14. British Patent 313,912 (1928).
15. British Patent 387,353 (1933).
16. British Patent 427,494 (1934).
17. British Patent 455,742 (1936).
18. British Patent 590,816 (1947).
19. R. E. Burk, H. Thompson, et al., *Polymerization*, Reinhold, New York (1937).
20. R. E. Burk and O. Grummitt, *The Chemistry of Large Molecules*, Interscience, New York (1943).
21. R. E. Burk, L. Taskowski, and H. R. Lakelma, *J. Am. Chem. Soc.* **63**, 3248 (1941).
22. R. E. Burk, B. G. Baldwin and C. H. Whitacre, *Ind. Eng. Chem.* **29**, 326 (1937).
23. W. H. Carothers, *Chem. Rev.* **8**, 397 (1931).
24. W. H. Carothers, *Collected Papers*, Interscience, New York (1941).
25. W. H. Carothers and I. Williams, *J. Am. Chem. Soc.* **53**, 4203 (1931).
26. W. H. Carothers and G. J. Berchet, *ibid*, **55**, 2807 (1933).
27. W. H. Carothers, J. E. Kirby and A. M. Collins, *ibid.* **55**, 789 (1933).
28. S. C. Cohen, *ibid.* **67**, 17 (1945).
29. Conference on High Polymers, *Ann. N. Y. Acad. Sci.* **44**, 263 (1943).
30. A. C. Cuthbertson, *Can. J. Research*, **20**, 103 (1942).
31. F. C. Ellenwood, N. Kulik, and N. R. Gay, *Cornell Univ. Eng. Expt. Sta. Bull.* **30** (1942).
32. H. Fikentscher, *Z. Angew. Chem.* **51**, 433 (1938).
33. P. J. Flory, *J. Am. Chem. Soc.* **59**, 241 (1937).
34. S. G. Foord, *J. Chem. Soc.* (London), **48** (1940).
35. J. Franck and E. Rabinowitch, *Trans. Faraday Soc.* **30**, 120 (1934).
36. V. J. Frilette, Abstracts of the 108th Meeting, *Am. Chem. Soc.* New York (September, 1945).
37. C. F. Frying and E. W. Harrington, *Ind. Eng. Chem.* **36**, 114 (1944).
38. G. Gee, C. B. Davis, and W. H. Melville, *Trans. Faraday Soc.* **35**, 1298 (1939).
39. H. Gelissen and P. H. Hermans, *Ber.* **58**, 285 (1925).
40. R. Ginell and R. Simha, *J. Am. Chem. Soc.* **65**, 705, 715 (1943).

41. G. Goldfinger, I. Skeist, and H. Mark, *J. Phys. Chem.* **47**, 578 (1943).
42. K. Gottlob, US Patent 1,149,577 (1915).
43. R. A. Gregg and F. R. Mayo, Am. Chem. Soc. Meeting (April 1946).
44. J. M. Grim, Mellon Institute of Industrial Research, Private Communication (1948).
45. E. Guth and O. Gold, *Phys. Rev.* **53**, 322 (1938).
46. W. D. Harkins, *J. Am. Chem. Soc.* **69**, 1428 (1947).
47. W. D. Harkins, *J. Chem. Phys.* **13**, 381 (1945).
48. W. D. Harkins, *ibid.* **14**, 47 (1946).
49. W. D. Harkins and R. S. Stearns, *ibid.* **14**, 215 (1946).
50. W. D. Harkins, R. W. Mattoon, and M. L. Corrin, *J. Am. Chem. Soc.* **68**, 220 (1946).
51. K. Hess and J. Gundermann, *Ber.* **70**, 1800 (1937).
52. K. Hess and W. Philippoff, *ibid.* **70**, 1808 (1937).
53. K. Hess, W. Philippoff, and H. Kiessig, *Naturwiss.* **26**, 184 (1938).
54. F. Hofman and K. Delbrück, German Patent 250,690 (1909).
55. F. Hopff and W. Kern, *Modern Plastics* **23**, 153 (1946).
56. H. Hulburt, R. A. Harman, A. Tobolsky, and H. Eyring, *Ann. N. Y. Acad. Sci.* **44**, 371 (1943).
57. W. H. Hunter and R. V. Yohe, *J. Am. Chem. Soc.* **55**, 1248 (1933).
58. E. P. Irany, *ibid.* **62**, 2690 (1940).
59. D. Josefowitz and H. Mark, *Polymer Bull.* **1**, 140 (1945).
60. S. Kamenskaya and S. Medvehev, *Acta Physicochim*, USSR **13**, 565 (1940).
61. W. Kern and H. Kammerer, *J. Prakt. Chem.* **161**, 81 (1942).
62. M. S. Kharasch, *J. Am. Chem. Soc.* **66**, 1438 (1944).
63. M. S. Kharasch, E. M. May, and F. R. Mayo, *J. Org. Chem.* **3**, 174 (1938).
64. M. S. Kharasch, A. T. Read, and F. R. Mayo, *J. Soc. Chem. Ind.* **57**, 752 (1938).
65. I. M. Kolthoff and W. J. Dale, *J. Am. Chem. Soc.* **67**, 1672 (1945).
66. P. L. Kooijman and W. L. Ghijseu, *Rec. Trav. Chim.* **66**, 247 (1947).
67. A. Kronstein, *Ber.* **35**, 4150 (1902).
68. H. Mark, *The Chemistry of Large Molecules*, Interscience, New York (1943).
69. H. Mark and R. Raff, *High Polymeric Reactions*, Interscience, New York (1941).
70. H. Mark and R. Raff, *Z. Physik. Chem.* **31**, 275 (1936).
71. C. S. Marvel and E. C. Horning, "Synthetic Polymers," in H. Gilman, *Organic Chemistry*, Wiley, New York (1943).
72. W. C. Mast, Eastern Regional Research Lab., Private Communication (1943).
73. M. S. Matheson, *J. Chem. Phys.* **13**, 584 (1945).
74. F. R. Mayo, *J. Am. Chem. Soc.* **65**, 2328 (1943).
74a. F. R. Mayo, and F. M. Lewis, *J. Am. Chem. Soc.* **66**, 1594 (1944).
75. J. W. McBain and A. M. Soldate, *J. Am. Chem. Soc.* **64**, 1556 (1942).
76. J. W. McBain, *Trans. Faraday Soc.* **9**, 99 (1913).
77. J. W. McBain, in *Advances in Colloid Science*, Vol. I, Interscience, New York (1942).
78. J. H. McClure, R. E. Robertson, and A. C. Cuthbertson, *Can. J. Research*, **B20**, 103 (1942).
79. S. Medvedev and P. Tseitlin, *J. Phys. Chem.* **49**, 211 (1945).
80. S. Medvedev and P. Tseitlin, *Acta Physicochim.* USSR, **20**, 3 (1945).
81. R. Mesrobian and A. Tobolsky, *J. Am. Chem. Soc.* **67**, 785 (1945).
82. H. W. Melville, *Proc. Roy. Soc.* (London), **A163**, 511 (1937).
83. R. G. Mulliken, *Rev. of Mod. Phys.* **14**, 265 (1942).
84. R. W. G. Norrish and E. F. Brockman, *Proc. Roy Soc.* (London), **A163**,

205 (1937).
85. H. F. Pfann, D. J. Solley, and H. Mark, *J. Am. Chem. Soc.* **66**, 983 (1944).
86. C. C. Price, et al., *ibid.* **64**, 1103, 2508 (1942); **65**, 757, 2380 (1943).
87. C. C. Price and C. E. Adams, *ibid.* **67**, 1674 (1945).
88. C. C. Price, *Ann. N. Y. Acad. Sci.* **44**, 368 (1943).
89. C. C. Price, *ibid.* **44**, 352 (1943).
90. C. C. Price and F. Kell, *J. Am. Chem. Soc.* **63**, 2798 (1941).
91. G. V. Schultz, A. Dinglinger, and E. Husemann, *Z. Physik. Chem.* **B43**, 385 (1939).
92. G. V. Schultz and F. Blaschke, *ibid.* **B51**, 75 (1942).
93. G. V. Schultz and E. Husemann, *ibid.* **B39**, 246 (1938).
94. G. V. Schultz, *Kunststoffe*, **33**, 224 (1943).
95. G. V. Schultz, *Z. Physik. Chem.* **A179**, 36 (1937).
96. G. V. Schultz, *ibid.* **A176**, 317 (1936).
97. E. Simon, *Ann.* **31**, 287 (1839).
98. H. Staudinger, *Die hochmolekularen organischen Verbindungen*, J. Springer, Berlin (1932).
99. H. Staudinger, *Trans. Faraday Soc.* **32**, 97 (1936).
100. H. Staudinger, *Ber.* **53**, 1081 (1920).
101. H. Staudinger, *Ann.* **447**, 110 (1926).
102. H. Staudinger, *Ber.* **59**, 3031 (1926).
103. H. Staudinger and E. Husemann, *Ber.* **68**, 1691 (1935).
104. Stobbe and Posnjak, *Ann.* **371**, 259 (1910).
105. E. Trommsdorf, *Chemie und Technologie der Kunststoffe*, Akademische Verlagsgesellschaft, Leipzig (1939) p. 320.
106. U. S. Patens 2,108,044 (1938) and 2,194,334 (1940).
107. U. S. Patent 1,775,882 (1930).
108. U. S. Patent 1,890,060 (1932).
109. U. S. Patent 1,976,224 (1934).
110. U. S. Patent 2,085,490 (1937).
111. U. S. Patent 1,881,282 (1932).
112. J. R. Vinograd, L. L. Fong, and H. M. Sawyer, Abstracts of the 108th Meeting, Am. Chem. Soc. New York (1945).
113. R. H. Wagner and T. Jauregg, *Ber.* **63**, 3213 (1930).
114. O. J. Walker and G. G. E. Wild, *J. Chem. Soc.* (London), 1132 (1937).
115. C. Walling, *J. Am. Chem. Soc.* **66**, 1602 (1944).
116. C. Walling, Abstracts of the 109th Meeting, Am. Chem. Soc., Atlantic City (1946).
116a. C. Walling, E. R. Briggs, and F. R. Mayo, *J. Am. Chem. Soc.* **68**, 1145 (1946).
117. G. S. Whitby, *Trans. Faraday Soc.* **32**, 315 (1936).
118. G. S. Whitby and R. M. Crozier, *Can. J. Research*, **6**, 203 (1932).
119. G. S. Whitby and M. Katz, *J. Am. Chem. Soc.* **50**, 1160 (1928).
120. R. Willstätter and J. Bruce, *Ber.* **40**, 3994 (1908).
121. R. York and E. F. White, *Trans. Am. Inst. Chem. Engrs.* **40**, 227 (1944).
122. L. Pauling, *The Nature of the Chemical Bond*, Cornell University Press, New York (1945).
123. A. E. Remik, *Electronic Interpretations of Organic Chemistry*, Wiley, New York (1943).
124. C. C. Price, *Reactions at Carbon-Carbon Double Bonds*, Interscience, New York (1946).
125. C. Ellis, *The Chemistry of Synthetic Resin*, Reinhold, New York (1935).
126. W. P. Hohenstein and H. Mark, *J. Polym. Sci.* **1**, 127 (1946).

127. I. M. Kolthoff and F. A. Bovey, *J. Am. Chem. Soc.* **70**, 191 (1948).
128. T. Alfrey, A. Bartovies, and H. Mark, *ibid.* **65**, 2319 (1943).
129. P. J. Flory, *ibid.* **65**, 372 (1943).
130. C. C. Price and D. A. Durham, *ibid.* **64**, 2508 (1942).
131. J. W. McBain, in E. O. Kraemer, *Advances in Colloid Science*, Vol. I, Interscience, p. 99-142, New York (1942).
132. V. J. Frilette and W. P. Hohenstein, *J. Polym. Sci.* **3**, 22 (1948).
133. R. Houwink, *J. Prakt. Chem.* **157**, 15 (1940).
134. W. Kuhn, *Kolloid-Z.* **62**, 269 (1933).
135. R. H. Ewart, in E. O. Kraemer, *Advances in Colloid Science*, Vol. II, Interscience, New York (1946) p. 197-251.
136. C. C. Winding, *Ind. Eng. Chem.* **40**, 1643 (1948).
137. Johnson and McEwen, *J. Am. Chem. Soc.* **48**, 469 (1926).
138. D. F. Straubel, Dow Chemical Co., Private Communication (October 1947).
139. U. S. Patent 2,241,770 (1941).
140. A. I. Goldberg, W. P. Hohenstein, and H. Mark, *J. Polym. Sci.* **2**, 503 (1947).
141. W. P. Hohenstein and H. Mark, *J. Polym. Sci.* **1**, 569 (1946).
142. H. Mark and R. Raff, *High Polymeric Reactions*, Interscience, New York (1943) p. 23.
143. K. H. Meyer, *Natural and Synthetic High Polymers*, Interscience, New York (1942), p. 12-26.
144. P. J. Flory, *J. Chem. Phys.* **10**, 51 (1942).
145. M. L. Huggins, *J. Am. Chem. Soc.* **64**, 1712 (1942).
146. D. M. French and R. H. Ewart, *Anal. Chem.* **19**, 165 (1947).
147. R. M. Fuoss and D. J. Mead, *J. Phys. Chem.* **47**, 59 (1943).
148. H. Mark, *Physical Chemistry of High Polymeric Systems*, Intersience, New York (1940) p. 258-296.
149. G. B. Jeffery, *Proc. Roy. Soc.* (London), **A102**, 161 (1923).
150. R. Simha, *J. Phys. Chem.* **44**, 25 (1940).
151. M. L. Huggins, *ibid.* **42**, 911 (1938); **43**, 439 (1939).
152. G. Gee, *Rubber Chem. and Tech.* **17**, 653 (1944).
153. H. Eilers, *Kolloid-Z.* **102**, 154 (1943).
154. E. O. Kraemer, *Ind. Eng. Chem.* **30**, 1200 (1938).
155. A. F. Martin, 103rd Meeting of the Am. Chem. Soc. (April 1942).
156. F. Howlett, E. Minshall, and A. R. Urquhart, *J. Textile Inst.*, **35**, 133 T. (1944).
157. S. Coppick, *Paper Trade J.* **119**, 26, 36 (1944).
158. R. S. Spencer and R. F. Boyer, *Polym. Bull.* **1**, 129 (1945).
159. K. B. Meyer, *Kolloid-Z.* **95**, 70 (1941).
160. M. L. Huggins, *Ind. Eng. Chem.* **35**, 980 (1943).
161. H. Mark, *Der feste Körper*, Hirzel, Leipzig (1938) p. 103.
162. G. V. Schulz and F. Dinglinger, *J. Prakt. Chem.* **158**, 136 (1941).
163. A. Bartovics and H. Mark, *J. Am. Chem. Soc.* **65**, 1901 (1943).
164. A. R. Kemp and H. Peters, *Ind. Eng. Chem.* **34**, 1097 (1942).
165. A. R. Kemp and H. Peters, *ibid.* **34**, 1192 (1942).
166. W. D. Baker, C. S. Fuller, and J. H. Heiss, *J. Am. Chem. Soc.* **63**, 2142 (1941).
167. R. Signer and H. Gross, *Helv. Chim. Acta.* **17**, 335 (1934).
168. P. M. Doty, B. H. Zimm, and H. Mark, *J. Chem. Phys.* **12**, 144 (1944).
169. H. Mark, *Ind. Eng. Chem.* **34**, 1343 (1942).
170. F. T. Wall, *J. Am. Chem. Soc.* **67**, 1929 (1945).
171. V. L. Simril, *Symposium on Plasticizers*, Interscience, New York (1947).
172. H. A. Winkelmann, *India Rubber World*, **113**, 799 (1946).

173. *Symposium of Plasticizers*, Interscience, New York (1947).
174. H. Mark, *Chem. and Eng. News*, **27,** 138 (1949).
175. C. F. Fryling, S. H. Landes, W. M. St. John, and C. A. Uraneck, *Ind. Eng. Chem.* **41,** 986 (1949).
176. C. F. Frying, *Ind. Eng. Chem.* **40,** 928 (1948).
177. I. M. Kolthoff and W. E. Harris, *J. Polymer Sci.* **2,** 41 (1947).
178. W. A. Schulze and W. W. Crouch, *Ind. Eng. Chem.* **40,** 151 (1948).
179. E. J. Meehan, *J. Polym. Sci.* **1,** 318 (1946).
180. W. C. Mast and C. H. Fisher, *Ind. Eng. Chem.* **41,** 790 (1949).
181. H. R. Snyder, J. M. Stewart, R. E. Allen, and R. J. Dearborn, *J. Am. Chem. Soc.* **68.** 1422 (1946).
182. F. T. Wall, F. W. Banes, and G. D. Sands, *J. Am. Chem. Soc.* **68,** 1429 (1946).
183. Plasticizer Symposium, 114th Meeting of the Am. Chem. Soc. *Ind. Eng. Chem.* **41,** 663 (1949).

chapter 3

Rheology of Paper Coatings and Instruments for the Measurement of Their Flow Properties

PERCY E. PIERCE

INTRODUCTION

Rheology is the science that deals with the deformation and flow of matter, common to most industrial processes; hence, a knowledge of rheology is essential for successful operation and control. The rheological behavior of matter has been the subject of extensive study

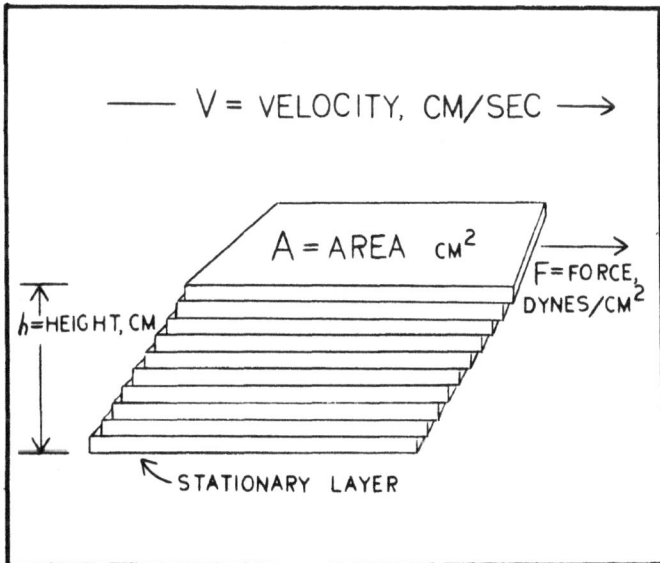

Fig. 3-1. Simple shear deformation.

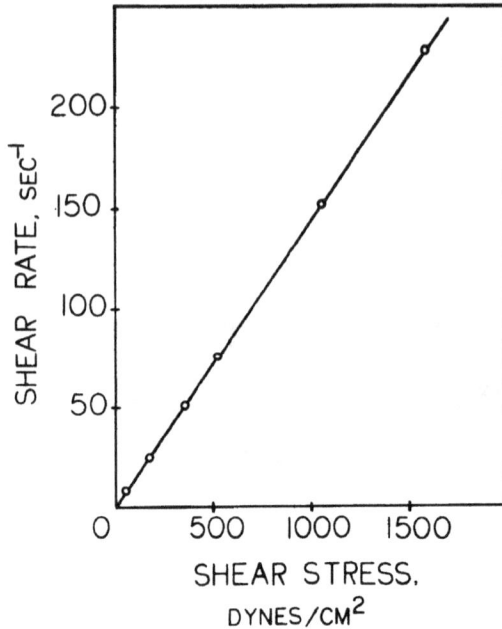

Fig. 3-2. Newtonian flow curve for a 40% solution of an acrylic polymer in toluene at 25°C.

and much of the rheological information of importance in paper coatings, inks, and related materials is summarized in specialized monographs.[1-5] In this chapter, the flow behavior of some typical materials is examined and a number of instruments useful for the investigation and control of coating rheology are described.

To investigate the rheological behavior of matter one must apply forces to a body and observe the resulting deformation of the system. A deformation is defined as the process of changing the relative position of the various parts of a body. Some deformations are reversible in that the work expended to deform the body is recovered when the body is returned to its original configuration. In other deformations, the energy supplied to produce the deformation is converted to heat; these irreversible deformations are called *flow*. The chapter deals with only the latter type of deformation, although in many materials both types of deformation occur simultaneously.

One of the common types of deformation is simple shearing, described in Figure 3-1. Simple shear can be pictured as a process in which infinitely thin parallel planes slide over each other in much the

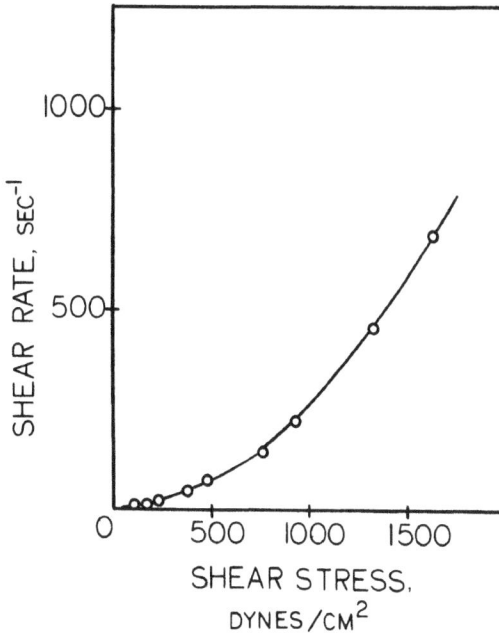

Fig. 3-3. Pseudoplastic flow curve for a latex coating.

same way as a deck of cards is spread. When steady flow occurs, the planes move at constant velocity. In Figure 3-1, if V is the velocity of the top plane with respect to the bottom plane located at a distance h, the rate of shear γ is equal to

$$\gamma = \frac{V}{h} \qquad (3\text{-}1)$$

The units of γ are 1/sec.

Because the energy of motion of a viscous fluid is being dissipated continuously and converted to heat, a force must be exerted continuously to maintain a steady flow. The shearing stress τ is defined as the ratio of the shearing force F to the sheared area A:

$$\tau = \frac{F}{A} \qquad (3\text{-}2)$$

The units of τ are dynes/cm².

The viscosity of the viscous fluid is defined as the ratio of the shear stress and shear rate:

$$\eta = \frac{\tau}{\gamma} \qquad (3\text{-}3)$$

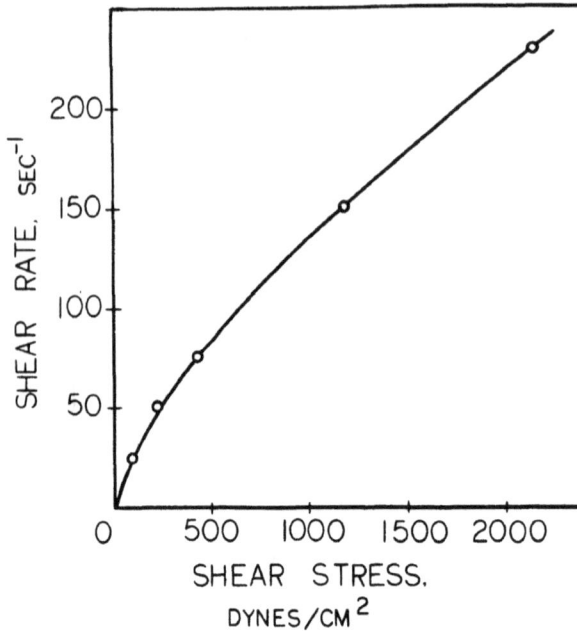

Fig. 3-4. Dilatant flow for a 70% clay suspension.

The unit of viscosity is dyne · sec/cm², called a poise.

When materials such as simple liquids, polymer melts, and suspensions are deformed, the shear rate is often a function of the shear stress, a dependence which can be shown by a graph of the two quantities. The resulting curve is called the flow curve or consistency curve of the substance.

FLOW CURVES

Most of the rheological data necessary for design and control of processes is contained in the flow curve. Certain types of flow curve or flow behavior are observed so frequently that they have been given names. Figures 3-2, 3-3, and 3-4 show three common types of flow curve exhibited by coatings and related materials.

Figure 3-2 shows *Newtonian* flow behavior, exhibited by single phase liquids composed of small molecules. Low molecular weight polymers and prepolymers, as well as sufficiently dilute polymer solutions, often show Newtonian flow. The data for 40% polymer solution given in Figure 3-2 show that low molecular weight polymers in solution can exhibit Newtonian flow behavior even at moderate concentrations.

Figure 3-3 shows an example of *pseudoplastic* flow, exhibited by many coatings. Polymer melts and solutions of high molecular weight polymers, as well as colloidal solutions, are frequently pseudoplastic. The degree of pseudoplastic behavior depends on the concentration of colloid or polymer, being greater at higher concentrations.

Figure 3-4 is an example of *dilatant* flow; this type of flow curve is generally observed in concentrated suspensions. The data shown in Figure 3-4 are representative of the type of dilatant flow encountered with clay suspensions. The degree of dilatant behavior increases as the concentration of suspended clay increases.

THIXOTROPY

The type of flow behavior discussed in the preceding section is independent of time, i.e., the value of shear stress developed in the material as a result of shearing at a given value of the shear rate does not depend on how long the material is sheared. Many coatings and

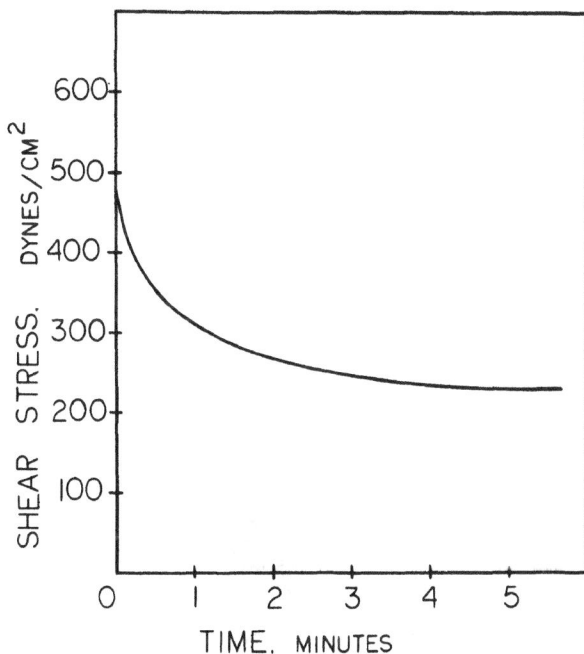

Fig. 3-5. Decay of the shear stress of a thixotropic coating as a function of time. $\gamma = 8.5$ sec^{-1}.

colloidal systems exhibit time-dependent flow effects; the most commonly encountered behavior is shown in Figure 3-5. The sample was sheared at a constant rate and the corresponding shear stress measured as a function of time. The initial value of the shear stress decayed over a period of five minutes to a final or equilibrium value slightly less than half the initial value. This material, and others like it that show a drop in shear stress with increasing flow time, are said to exhibit thixotropy.

After the material has been sheared to a steady shear-stress value and allowed to rest, most thixotropic substances recover approximately their initial behavior. The recovery time may range from a minute or less to several days or even weeks, depending on the material. Figure 3-6 shows the shear stress recovery of the thixotropic coating whose decay data were shown in Figure 3-5. The rate of decay and recovery depend on the shear rate of the experiment. Figure 3-7 shows the initial and final, or equilibrium, values of shear

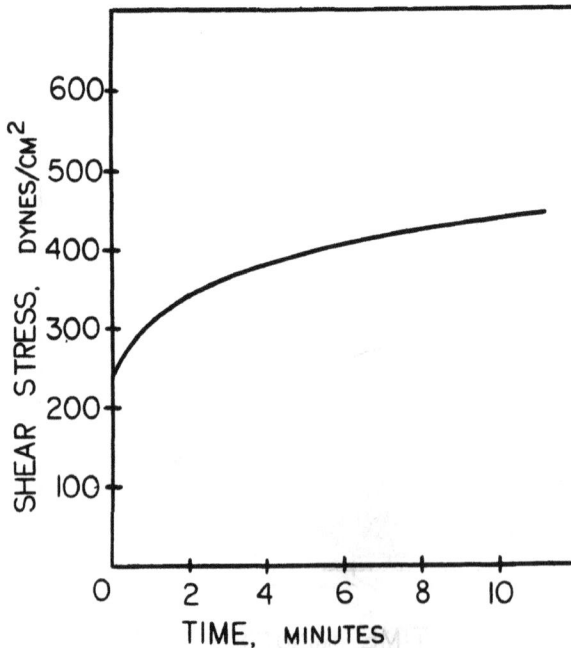

Fig. 3-6. Rebuilding of the shear stress of a thixotropic coating as a function of time. $\gamma = 8.5$ sec^{-1}.

stress plotted logarithmically against the corresponding shear rates at which they were determined. These data are the counterparts of the flow curves for substances that do not show time-dependent effects. The kinetic recovery and decay data taken over the range of shear rate of interest give a complete specification of the thixotropic flow behavior.

Many thixotropic substances show a yield value in the undisturbed state, i.e., a certain minimum force or stress is required to produce flow. Once the sample has been sheared, the material flows at lower stress levels until the original sample structure is recovered. The data shown in Figure 3-7 illustrate this behavior. The downturn of the initial shear stress curve indicates that in the undisturbed state a stress of approximately 500 dynes/cm^2 is required to produce flow. In the final or equilibrium sheared state, flow takes place at shear stress levels less than 200 dynes/cm^2. This behavior is characteristic of many thixotropic coatings.

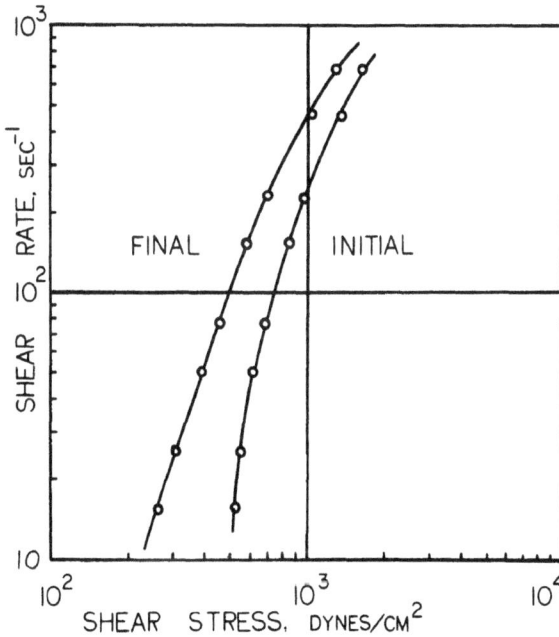

Fig. 3-7. Initial and final (or equilibrium) values of the shear stress for a thixotropic coating.

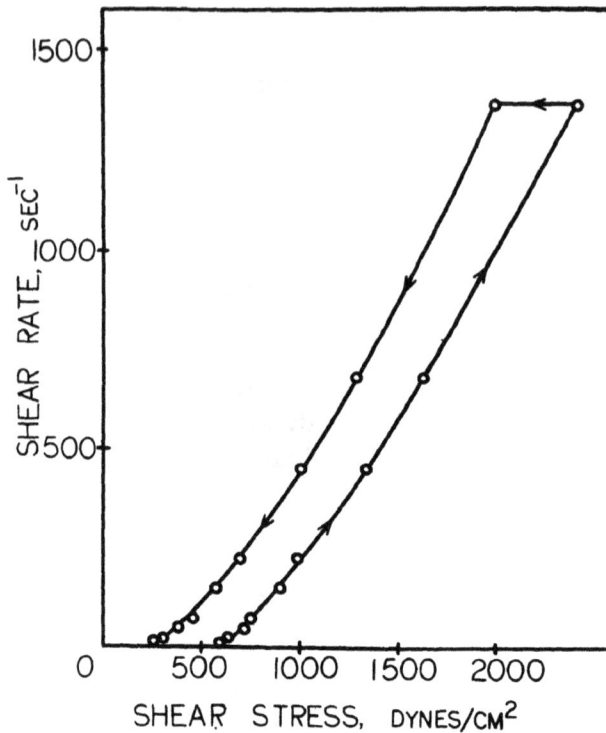

Fig. 3-8. Thixotropic loop for a coating.

Another method used to investigate time-dependent behavior involves running a thixotropic loop. Such loops can be determined in
a number of ways.[3,5,6] Figure 3-8 shows a thixotropic loop for the
coating whose decay and recovery data were just considered. The
loop was run by starting at a low rate of shear and recording the
value of shear stress developed by the material. The shear rate was
increased as rapidly as possible to the highest rate available; this
process is represented by the up portion of the loop. The sample was
then sheared at this rate to its equilibrium shear-stress value (flat
portion of the loop) and the shear rate decreased rapidly back to the
lowest value. This step is represented by the down portion of the
loop. The degree of thixotropy of the sample is related to the area
of the loop. The loop method defines only approximately the thixotropic behavior of a material because it does not yield information
on the rate of decay and rebuilding of the shear stress. It has the

advantage of being a rapid method for detecting time-dependent effects. Methods suitable for investigating thixotropic coatings are described by Van Wazar et al,[5] Pierce and Donegan,[6] and Doherty and Hurd.[7]

EFFECT OF TEMPERATURE AND CONCENTRATION ON VISCOSITY

Viscosity is defined as the ratio of the shear stress to shear rate. Figure 3-9 shows the viscosities of a pseudoplastic coating and a Newtonian oil plotted against shear rate. The viscosity of the pseudoplastic coating depends strongly on the shear rate; the viscosity of the Newtonian oil is independent. Many authors prefer to represent the rheological behavior of a material by plotting viscosity against shear rate or shear stress. This method of presenting the data is most widely used for systems that do not deviate too far from Newtonian behavior.

The rheological behavior of a material depends strongly on the

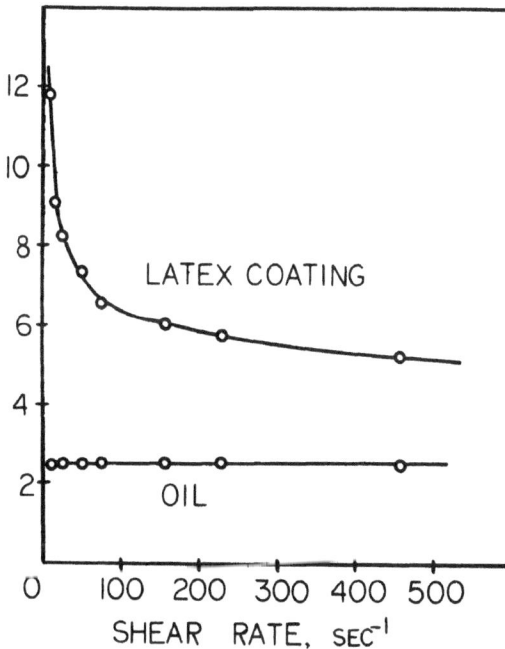

Fig. 3-9. Viscosity of a pseudoplastic latex coating and a Newtonian oil as a function of shear rate.

temperature; the viscosities of many materials decrease rapidly as the temperature is raised. Andrade's equation[8] describes approximately the temperature dependence of the viscosity η of many substances:

$$\eta = Ae^{\frac{B}{T}} \qquad\qquad (3\text{-}4)$$

where A and B are constants and T is the absolute temperature. Figure 3-10 shows a semilog plot of viscosity against the reciprocal of the absolute temperature, verifying the applicability of Andrade's equation.

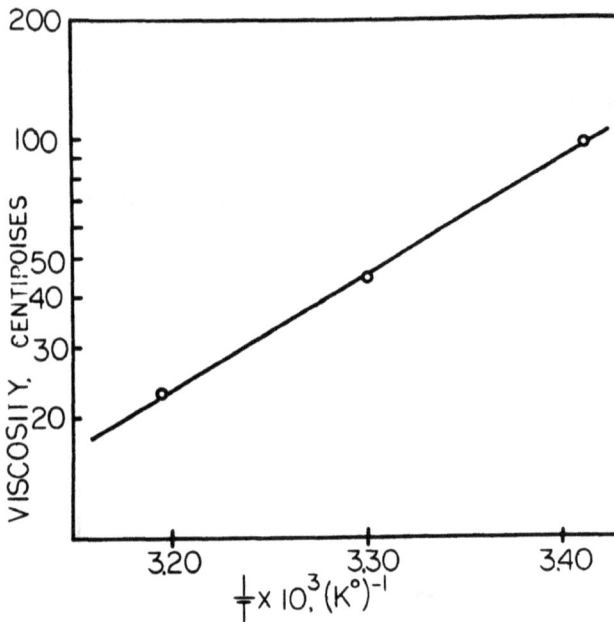

Fig. 3-10. Variation in viscosity of an oil as a function of temperature.

The concentration dependence of the viscosity of polymer solutions, colloids, and coatings is generally complicated. The net result of increasing the concentration of polymer or dispersed phase is to increase the viscosity of the system. Also, the tendency for non-Newtonian and thixotropic flow behavior is also increased at higher concentrations. No general equation adequately expresses the concentration dependence of these systems. Some of the equations proposed to represent the behavior of polymer solutions are reviewed by Patton[4] and by Streeter and Boyer.[9] The three-volume work edited by Eirich[1] has a number of useful articles dealing with the concentration and temperature dependence of the viscosity of colloid and polymer solutions.

DETERMINATION OF RHEOLOGICAL BEHAVIOR

The choice of instruments to measure the rheological properties of a material depends on the complexity of the flow behavior of the material to be tested, the motives of the user, and the nature of the processing the material receives. If a complete description of the rheology of a material is required over a wide range of shear rates, a rotational viscometer with electronic data recording is required. Such an instrument, properly used, can discriminate time-dependent effects and non-Newtonian flow. For control work or routine checking of the quality of a product, simpler devices such as bubble tubes or viscosity cups may be adequate. In general, the best quality-control methods check the rheological behavior of the material under conditions that represent processing. For example, it is poor technique to use a low shear-rate viscosity measurement to check the quality of a material that will be used under high shear-rate conditions, unless the material is Newtonian.

In this section, some of the more widely used instruments in the field of coatings are described. A description of most commercially available viscometers (with operating instructions) can be found in the book by Van Wazar et al.[5]

Rotational Viscometers

Figure 3-11 is a schematic diagram of a concentric cylinder viscometer. This type of instrument consists of a cylindrical bob of radius R_b, which is rotated with an angular velocity Ω inside a stationary cup of radius R_c. The sample is sheared between the sample cup and rotor. The rate of shear can be varied by varying the speed of the rotor. The shearing stress at the bob is measured by determining the retarding or frictional torque exerted on the rotor. The shear stress is related to the torque by the relationship

$$\tau = \frac{M}{2\pi} R_b^2 h \tag{3-5}$$

where M is the torque, R_b is the radius of the rotor, and h is the distance the rotor is immersed in the sample.

The angular velocity Ω of the rotor is related to the shear rate by

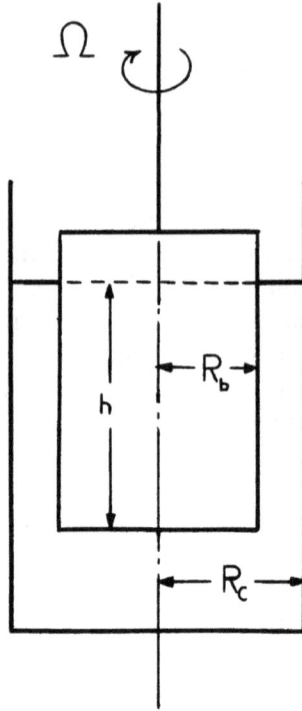

Fig. 3-11. Schematic diagram of a concentric cylinder viscometer.

the equation

$$\Omega = \frac{1}{2} \int_{\tau_b}^{\tau_c} \gamma\tau \frac{d\tau}{\tau} \qquad (3\text{-}6)$$

where τ is the shear stress, γ is the shear rate, and τ_c and τ_b are the shear stresses at the cup and bob, respectively. Generally, τ_b and Ω are measured experimentally. The rate of shear at the bob γ_b is required in order to construct the flow curve. If the shear rate and shear stress obey the power flow law

$$\gamma = k\tau^N \qquad (3\text{-}7)$$

then equation (3-6) can be solved,[5,6,10] giving

$$\gamma_b = \frac{2\Omega}{1-\varepsilon^{-2}}\left[1 + (N-1)\ln\,\varepsilon + \frac{(N-1)(N-2)}{3}(\ln\,\varepsilon)^2 + \ldots\right] \qquad (3\text{-}8)$$

where $\varepsilon = R_c/R_b$ and $N = d\ln\Omega/d\ln\tau_b$. If ε is sufficiently close to unity, terms beyond the first or second in the series for γ_b are seldom required. In applied work, the higher correction terms are often omitted and the first, or Newtonian, term is used. This practice can lead to serious errors if ε is large.

The Rotovisco viscometer (Figure 3-12) manufactured by the Haake Company and distributed in the United States by the Polyscience Corporation has been found to be a satisfactory instrument for the investigation of coatings. The instrument is quite versatile and can be purchased with attachments to cover a shear-rate range from 10^{-2} to $1.4 \cdot 10^4$ sec^{-1}. The instrument is suitable for measuring time-dependent effects if equipped with an electronic recorder and is described with other commercially available concentric cylinder viscometers by Van Wazar et al.[5]

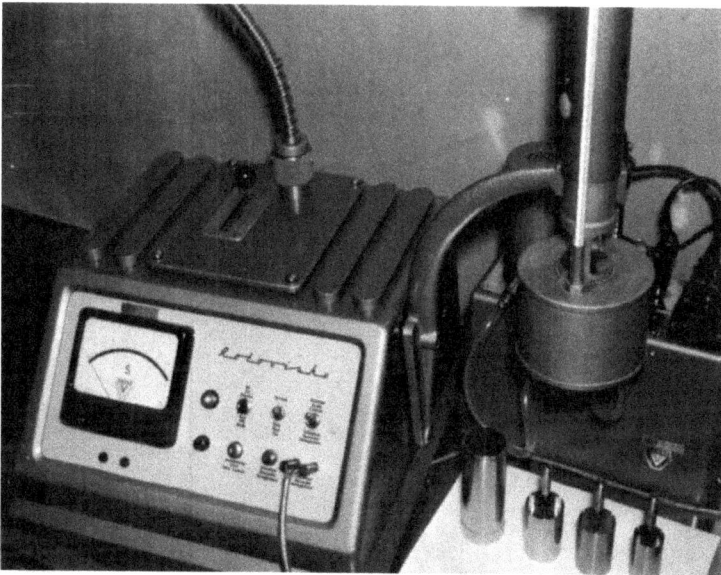

Fig. 3-12. Roto viscometer and the MV series concentric cylinder geometry rotors and cup. (*Courtesy of The Glidden Co.*).

Another type of widely-used rotational viscometer is the cone-and-plate viscometer. Figure 3-13 is a schematic diagram of the cone-and-plate geometry. The appratus consists of a cone of small angle α, rotating with an angular velocity Ω on a flat plate. The shear stress is determined by measuring the viscous retarding torque exerted on the rotor by the sample, which is between the cone and plate. The shear stress is related to the retarding torque M by the

equation

$$\tau = \frac{3M}{2\pi R^2} \qquad (3\text{-}9)$$

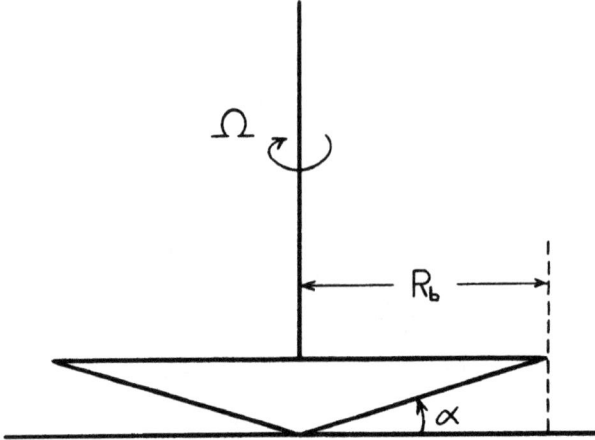

Fig. 3-13. Schematic diagram of a cone-and-plate viscometer.

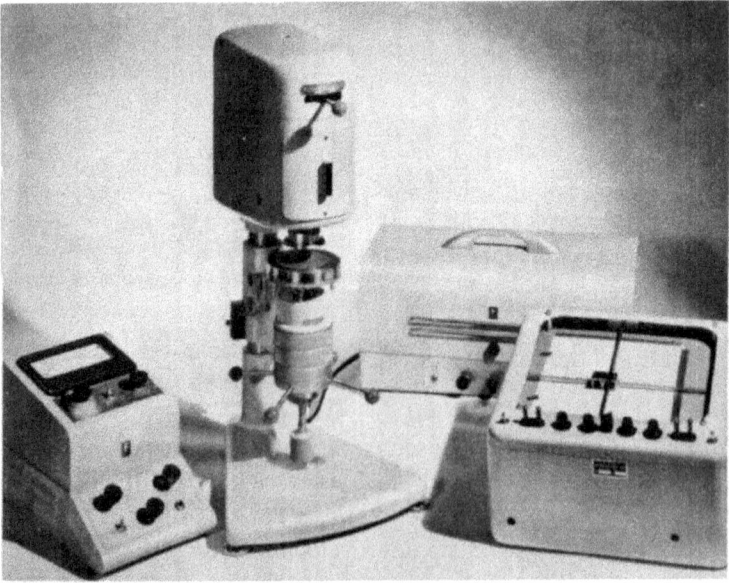

Fig. 3-14. Ferranti-Shirley cone-and-plate viscometer with an automatic program unit and X-Y recorder (*Courtesy of Ferranti Electric Inc.*).

where R is the radius of the cone. The rate of shear is related to the angular velocity and angle of the cone by the relation

$$\gamma = \frac{\Omega}{a} \qquad (3\text{-}10)$$

The popularity of the cone and plate viscometer is due in part to the simple equations relating shear rate and shear stress to experimental quantities without the necessity for elaborate calculation in the case of non-Newtonian materials. A cone-and-plate viscometer with electronic recording apparatus is well suited for studying time-dependent effects, but is not as suitable for measuring coarse suspensions as the concentric cylinder viscometer.

The Ferranti–Shirley viscometer (Figure 3-14) is probably the most widely used cone-and-plate viscometer commercially available.

Fig. 3-15. Stormer viscometer with the paddle wheel rotor of the type used in many coating laboratories. The above model has a stroboscopic attachment for easier measurement (*Courtesy of The Glidden Co.*).

This instrument has a rate of shear range from 0 to $1.8 \cdot 10^4$ sec^{-1}. It is available with a control unit and X-Y recorder that makes it possible to run thixotropic loops automatically. The operation and design of this instrument is described by Van Wazar et al.[5]

Rotational viscometers of less elaborate design are used for routine laboratory work. The Stormer viscometer shown in Figure 3-15 is widely used in coating laboratories. Several models of the Stormer design are in use, differing in the type of rotating cylinder or paddle-and-cup geometry. None of the standard modifications are amenable to mathematical treatment, so that the instrument must be calibrated with standard oils. The time required for a definite number of rotor revolutions (generally 100) when immersed in a fluid and driven with a definite weight is a measure of the viscosity of the liquid. The

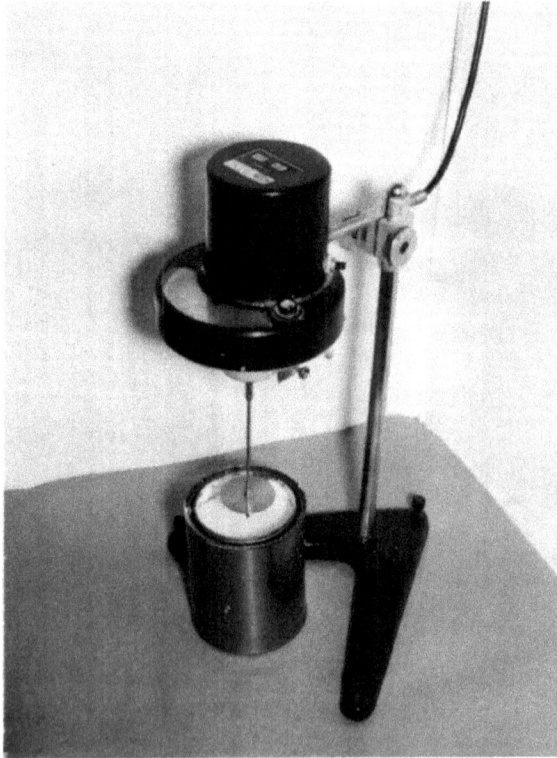

Fig. 3-16. Brookfield viscometer with a typical spindle rotor. This instrument is particularly convenient for routine viscosity measurements (*Courtesy of The Glidden Co.*).

shear stress may be varied by changing the falling weights that drive the instrument. A flow curve can be prepared by plotting rotational velocity against applied torque. The paddle version of the Stormer viscometer, specified in ASTM test method D562-55, is the accepted standard instrument of the paint industry. The loading necessary to maintain a speed of 200 rpm is determined, and converted to Krebs units (KU) by an arbitrary scale.[4]

The Brookfield viscometer shown in Figure 3-16 is a popular instrument in industrial laboratories. The spindle, which is immersed in the fluid to be tested, is driven by a synchronous motor through a multispeed gear box. Several available models of the Brookfield viscometer differ in the types of spindle and speed range. The retarding viscous torque is sensed by a spring and is indicated on a calibrated dial. The apparent viscosity is found from a table supplied with the instrument, which converts spindle number, rotational velocity, and dial reading to viscosity.

This insrument has a number of useful features: The various models permit a wide range of viscosities to be determined; it is convenient to use and to clean; and many special attachments are available. The ordinary attachments such as the disk-type spindles are not amenable to mathematical analysis because of large end effects. The Brookfield viscometer is specified in ASTM test method D2196-63T for the investigation of the rheological properties of non-Newtonian materials.

Capillary Viscometers

Glass capillary viscometers such as the Ostwald viscometer shown in Figure 3-17 are widely used to measure the viscosity of clear Newtonian liquids or solutions. The viscometer is filled with a known volume of liquid, placed in a constant-temperature bath, and after temperature equilibrium has been attained, the liquid is drawn into the calibrated bulb. The time required for the liquid to flow under gravity past the initial and final marks is taken, and the viscosity is determined from the equation

$$\eta = A \varrho t \qquad (3\text{-}11)$$

where A is the calibration constant of the instrument, ϱ is the density of the liquid, and t is the flow time. A kinetic energy correction is required for times less than about 200 seconds.[5] Instruments of this type are widely used to determine the intrinsic viscosity of polymers.[11]

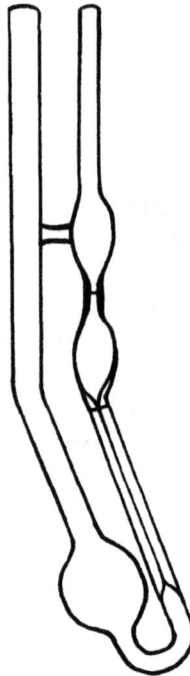

Fig. 3-17. Cannon-Fenske Viscometer—an improved version of the Ostwald viscometer.

Efflux Viscometers

Efflux viscometers or viscosity cups consist of a container with a standardized opening. Viscosity is measured by determining the length of time required for the liquid to flow from the filled cup. Because the length of the capillary in the typical cup is short, end effects and turbulence make mathematical analysis difficult.

These instruments are calibrated with liquids of known viscosity and are useful for comparison measurements. They are popular in the coatings industry because of their simple operation and low cost. Many different varieties of cup are in everyday use in this industry; typical of these are Ford cups (ASTM D1200-58) and Zahn cups. Gardner[2] describes the various types in some detail.

Falling Ball and Rising Bubble Viscometers

If a rigid spherical body falls in a viscous medium, it attains a constant velocity. Stokes' law relates the velocity of fall to the viscosity of the medium:

$$\eta = 2(\varrho_b - \varrho) R_b^2 \frac{g}{9v} \qquad (3\text{-}12)$$

where η is the viscosity, ϱ_b the density of the ball, ϱ the density of the medium, R_b the radius of the ball, g the gravitational constant, and v the velocity of fall. For Stokes' law to apply, the velocity of fall must be below certain limits and the radius of the ball must be small compared with the dimensions of the container holding the fluid.[5] The ball is dropped into the container and the time of fall between two marks a known distance apart is determined. The viscosity can be calculated from Equation (3-12) if the conditions for the validity of Stokes' law are fulfilled. Often the instrument is calibrated with materials of known viscosity and the corrections for edge effects and velocity are ignored.

In the Hoeppler viscometer (Figure 3-18), a ball is rolled down

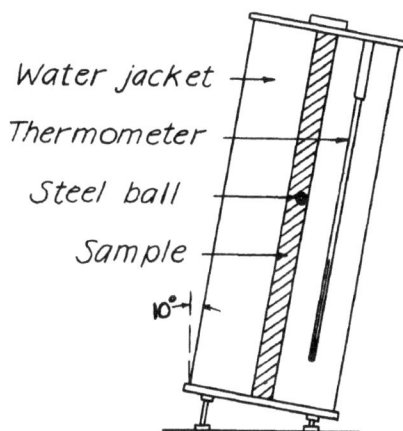

Fig. 3-18.   Hoeppler rolling-ball viscometer.

an inclined tube filled with the liquid whose viscosity is to be determined. The edge effects are complicated, so that calibration of the instrument with liquids of known viscosity is required. This viscometer gives precise results and is therefore accurate for control purposes.

The Gardner-Holdt bubble viscometer (Figure 3-19) is widely used in the paint industry. It consists of a set of standard tubes nearly filled with mineral oils of varying viscosity. The oil of unknown viscosity is placed in a standard tube, which is filled to the mark and

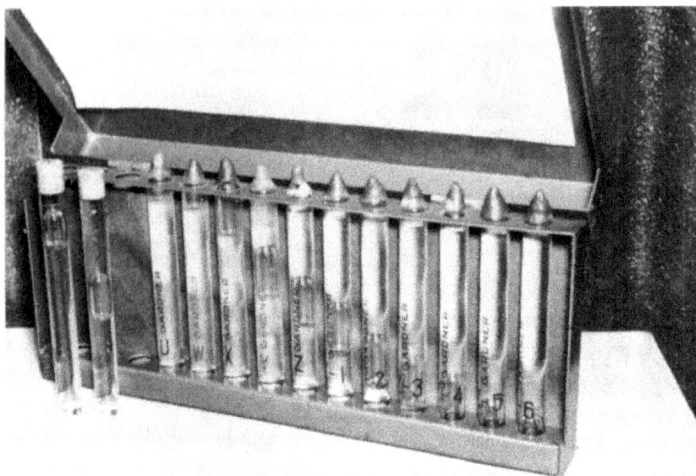

Fig. 3-19. Gardner-Holdt bubble-tube viscometer. Bubble-tube viscosity measurements are frequently used to characterize the viscosity of clear resin solutions (*Courtesy of The Glidden Co.*).

corked. The tubes are not completely filled so that a bubble is trapped in each. That containing the test oil is placed with the standard tubes in a rack, which is then a inverted. The bubbles in the tubes rise at different rates depending on the viscosities of the liquids they contain; the test oil is then matched with one of the standards.

Several sets of Gardner tubes covering different viscosity ranges are available. The method is used for clear or semiclear liquids with letter rating to indicate viscosity. An approximate conversion to ordinary viscosity units is possible.[4] The use of the bubble viscometer for determining the viscosity of resin solutions is discussed in ASTM test method D1725-62.

REFERENCES

1. F. R. Eirich, ed. *Rheology: Theory and Applications*, Academic Press, New York, Vol. 1 (1956); Vol. 2 (1958); Vol. 3 (1960); Vol. 4 in press.
2. H. A. Gardner and G. G. Sward, *Paint Testing Manual*, 12th Edition, Gardner Laboratory, Inc. Bethesda, Md. (1962).
3. H. Green, *Industrial Rheology and Rheological Structures*, Wiley, New York

(1949); Chapman and Hall, London (1949).

4. T. C. Patton, *Paint Flow and Pigment Dispersion*, Interscience, New York (1964).

5 J. R. Van Wazar, J. W. Lyons, K. Y. Kim, and R. E. Colwell, *Viscosity and Flow Measurements*, Interscience, New York (1963).

6. P. E. Pierce and V. A. Donegan, *J. Paint Tech.* **38**, No. 492, 1, (1966).

7. D. J. Doherty and R. Hurd, *J. Oil & Colour Chemists' Assoc.* **41**, 42 (1958).

8. E. N. Da C. Andrade, *Nature*, **125**, 309 (1930).

9. J. Streeter and R. F. Boyer, *Ind. Eng. Chem.* **43**, 1790 (1951).

10. I. M. Krieger and H. Elrod, *J. Appl. Phys.* **24**, 134 (1953).

11. F. W. Billmeyer, Jr., *Textbook of Polymer Chemistry*, Interscience, New York (1962).

chapter 4

Water-Base Coatings

THOMAS W. BUSCH

A water-base coating for printing or specialty coated papers consists of a solid phase, dissolved or dispersed in water. The solid film or layer that remains and adheres to the paper substrate when the water is removed may be composed of a combination of adhesives, inorganic minerals, or organic compounds. Adhesive coatings such as casein, animal glue, vegetable protein, starch, cellulose derivatives, and various synthetic polymers can be dissolved, colloidally dispersed, or emulsified in water. The particle size of the adhesive substance in water determines whether the system is a true solution or a colloidal suspension.

The solid is said to be soluble when—as molecules or ions—it diffuses throughout the water to form a homogeneous mixture as a single phase. On the other hand, suspensions and emulsions are two-phase systems with dispersed particles where the water acts as a dispersing medium rather than as a solvent. Singly, or in combination, adhesives form films or coating layers when applied to paper. In practice, these substances are blended with insoluble mineral pigments or other complex materials to provide a superior printing surface, to obtain a decorative effect, or to realize a specific functional value. A comprehensive technical knowledge of the chemical and physical properties of such coating materials is necessary to utilize these building stones and to arrange them properly to accomplish a specific result.

HISTORY

Coated paper was developed for commercial use about 1885 to fulfill a demand for better printing results. At the outset, a thin layer of animal glue and pigment was applied to the surface of a moving paper web and was then dried and calendered. At the turn of the century, casein began to replace hide glue as the principal adhesive.

The use of kaolin clay as the pigment was soon followed by calcium carbonate, barium sulfate, lithopone, satin white, talc, titanium dioxide, and alumina, each of which has certain desirable characteristics. The adhesive in the coating binds the pigment particles, one to another and to the base paper, to prevent fracture or removal of the coating pigment while printing. The pigments provide the necessary hiding power to cover the base paper and to make the surface uniformly receptive to ink.

Early in 1900, natural starches were available as an adhesive for paper coating. The raw starch product produced pastes that were too viscous to make coating mixtures with satisfactory flow properties. Chemical modification by treatment with calcium hypochlorite was introduced and used for many years. Coating starches are now modified by oxidizing treatment, dextrinization, enzymes, and partial chemical substitution such as ethylation. In one of these forms starches are used to a much greater extent than casein, particularly in large-volume publication grades.

Protein derived from soybeans was next used as a possible substitute for casein in coatings. The bonding strength of soybean protein is approximately equal to that of casein but it is somewhat darker in color and requires a strong alkali to develop its full adhesive strength.

Synthetic Adhesives

During the early 1940's, synthetic resin latices were introduced to improve gloss, flexibility, compressibility, and wet-rub resistance of the coating. Seldom used alone, when they are blended with other adhesives in appropriate amounts desired improvements can be obtained.

Other water- and alkali-soluble synthetic resins are also available. One of the most interesting is polyvinyl alcohol, which is not subject to putrefaction or degradation, and has a much greater adhesive strength than has casein.

Early Methods

At first, coated paper was produced by a separate conversion process independent of the paper-making machinery. Such a product filled a need for which uncoated paper was unsatisfactory. Since then, conversion coated paper has had to compete with vastly improved supercalendered plain paper and those coated papers that are produced by applying a thin coating at high speeds as an integral part of the paper machine. Initially, machine-coated paper was quite inferior to that made by the conversion process, but with the skill and experience developed in the past decade its quality is approaching that of good conversion grades.

Coating was first applied in a continuous operation to only one side of the sheet. Machines have been developed, on and off the paper machine, to coat both sides simultaneously. Additional progress has led to a continuous coating process by which it is possible to apply more than one layer of coating to one or both sides of the web.

UNIT PROCESSES

Five unit processes are required to produce coated paper: (1) preparation and blending of the coating mixture; (2) application to the paper; (3) drying the paper web; (4) calendering or finishing the surface; and (5) cutting or rewinding, inspecting, packing, and shipping.

Each unit process of the coating industry is vast in scope and is in a rapidly increasing state of development. These advances are characterized partly by the introduction of numerous synthetic compounds suitable for coating improved printing and specialty papers. Simultaneously, rapid progress in mixing, technology, application methods, and drying and finishing machinery has marked the last decade.

Extensive progress in this important and expanding industry resulted from the demand for uniformly high-quality coated paper for graphic arts and functional applications. The major objective of coated paper for the printing industry is to permit the reproduction of the original subject as closely as possible. Specialty coated papers perform a function in addition to, or in place of, that required for graphic arts reproduction. This chapter outlines the technology of materials and methods common to both objectives and places particular emphasis on the newer specialty coated paper materials and applications. The information deals primarily with the nature of coating processes for off-the-machine conversion.

THE OBJECT OF COATING

Base papers are coated to impart certain desirable properties which render the coated sheet useful for specific eventual applications.

The reasons for applying a surface coating or saturating a paper web are as numerous as the functions that coated paper can be designed to perform. Each coating layer, usually very thin compared with the thickness of the paper substrate, performs a definite task in the market place.

Properties of a coated surface that can be gainfully applied may be classified as physical, optical, chemical, thermal, or electrical. No single coating necessarily contributes specific application value simultaneously in all categories, but physical and optical characteristics are generally involved. Examples to illustrate practical utilization of these properties include:

1. Level surfaces and attractive hues;
2. Resistance to wear by abrasion, folding, and scuffing;
3. Resistance to chemicals, water, grease, solvents, air, and gases;
4. Functional surfaces that are heat, light, and electrically-sensitive, and chemically reactive;
5. Surfaces that remain free from odor, taste, mold, and bacteria.

Commercial Uses

Perhaps the earliest and most common reason for applying coating, white or colored, matte or glossy finish, is to provide a satisfactory printing surface. The coating layer must be smooth, continuous, dense, and ideally suited for the reception of printing inks.

A sheet of bare cellulose fibers, such as is used in most newsprint, is not well suited to high fidelity printing because of transparency of structure and irregularities of surface. Ordinary filled paper, properly calendered, lacks a perfection in surface smoothness for accurate reproduction of tiny ink dots in half-tone printing. The quality of a simple filled sheet is enhanced by coating its surface with a thin film of finely divided mineral pigment suspended in an adhesive mixture. When such a paper is calendered, the result is a smooth, even-textured surface with a microporosity that helps to control ink absorption and permits reproduction of fine half-tone engravings with remarkable accuracy.

Colored coated papers add a decorative dimension to printing applications and are also in demand for purely aesthetic reasons in gift wrapping and colorful advertising. Colored coatings demand pigments that are adequately moisture resistant to prevent bleeding when exposed to water. Resistances to dry and wet abrasion, as well as light fastness, are also important in fancy decorative papers.

The manufacture of paper boxes, trays, liners, and corrugated containers designed for packaging and display of greasy products was once dependent on lamination. Formerly, it was almost universal practice to laminate a layer of greaseproof glassine paper, foil, or plastic film to paperboard and thereby to obtain a satisfactory barrier against water vapor, grease, volatile flavor compounds, and atmospheric gases. This process had the serious disadvantage of high cost and difficult utilization of waste. The introduction of special functional coatings has replaced many of the laminating operations with water-base coating processes. The coating technique permits close regulation of film thickness and broadens the range of materials and base stocks that can be treated.

Label papers, soap wrappers, box coverings, and shelf and drawer linings require careful selection of coating ingredients to provide stability and freedom from mold and odor.

The coated surface must remain flexible, scuff-resistant, and withstand exposure to high humidity or wiping with a damp cloth.

Reproduction Papers

The office copy market, which utilizes vast quantities of special functional coated papers, has grown from virtually nothing in 1950 to a major industry for supplying coated paper. Thermography requires a sensitized paper, chemically constructed to develop an image by heat or by infraradiation. Electrostatics had almost no commercial application in 1960, but by 1965 it had became a market of over 100 million dollars. The special coatings utilized in electrofax type papers consist of a specially prepared zinc oxide, a binder resin, and a dye or dyes to provide panchromatic sensitivity. The coating is electrophotoconductive, the basis for image formation by means of electrical and light energy.

The engineering copy market is primarily concerned with the reproduction of engineering drawings by the diazo process. A diazonium compound is one containing two associated nitrogen atoms. A colored dye forms in the appropriate area when the diazo compound is joined to a coupler. In this process, both the diazo compound and a coupling agent are coated on paper.

Major reproduction processes involve multicopy duplication by means of special coated offset and spirit paper masters. Since 1950, the offset process, which uses direct image or transfer paper for lithographic masters, has become one of the most important reproduction processes.

Composition

The multiple-business-forms market is concerned with producing copies of sales transactions by obtaining replication through pressure, whether by writing or by typewriter impact. Although formerly accomplished exclusively by carbon-tissue interleaving, pressure-sensitive transfer and self-contained systems, which do not require carbon tissue, have been introduced. Special coated papers that involve a chemical reaction between two special coatings or the physical transfer of a clean colored coating layer from one sheet to another have been introduced since 1955.

Whatever the field of application, the coating formulation must fulfill a number of basic requirements. It must penetrate only enough to bond satisfactorily to the surface of the paper. It must have adequate flow to be spread freely at expected machine speeds and must level and set after application to the base paper. On drying, an acceptable surface that can survive the subsequent unit manufacturing processes is required. Frequently it is difficult to obtain all of the desired properties with a single formulation. Usually, a particular system represents the best possible compromise to fit the manufacturing equipment available, to satisfy the eventual requirements, to minimize the cost, and to maintain optimum efficiency.

COATING TECHNOLOGY

The science, or accumulation of organized knowledge, established by observation and experiment in the paper coating industry has grown rapidly since 1955. Demands for larger quantities of coated papers to perform more numerous functions have increased at an unexpected rate. These factors have placed great emphasis on the requirements for advanced technology to coat, dry, and finish coated papers at maximum production rates and optimum efficiency. The coating technologist has been urged to learn more about the abilities and limitations of coating materials and the behavior of fluid coating systems throughout the various unit processes of the coating operation.

Influence of Materials

The physical state, chemistry, and compatibility of individual pigments and adhesives, and their interrelationship when blended, determine the coating behavior during actual operation. In most instances it is necessary to formulate a coating system that produces optimum results with a given base paper and available coating equip-ment. Off-the-paper-machine coating equipment such as the air knife and two-side roll coaters operate near optimum with fluid or low viscosity coatings.

The ratio and kind of adhesives, insoluble pigments, dispersing agents, additives and the solid content of the coating mix are major factors that affect the viscosity or flow properties of a coating. Usually the pigment comprises 75-90% by weight of the total solids. Dry coating pigments consist of agglomerates and aggregates built up by the individual ultimate particles that must be dispersed by mechanical mixing. Reflocculation of these particles in water-base coatings is prevented by using dispersing agents, which are adsorbed on the pig-ment crystals and create electrical repulsion of the particles. This action assures a reduction in viscosity at a given percentage of solids or makes possible a higher solids content at an acceptably low viscosity.

Colloidally dispersed natural adhesives with low molecular weights yield low viscosity, whereas the high-molecular-weight materials result in higher viscosity. Synthetic latex adhesives, which are aqueous dispersions of water-insoluble polymers, produce low-viscosity high-solid coatings. Their performance in pigmented coatings is deter-mined by distribution of particle size rather than by molecular weight.

Flow Properties

The adhesive tends to dominate the flow properties of the coating. The relationship between the rate of flow (or fluidity) and the shearing force applied to the coating is usually used to describe the flow pro-perties of a liquid system. Forces may be applied or imposed on a coating liquid by pump, agitator, coater roll, or metering action on a trailing-blade or air-knife coater.

A knowledge of the flow properties enters into mixing of coatings and shearing action inherent in methods of application. Rheological changes occur during the short dwell-interval of coating and paper before drying. Migration or redistribution of coating components because of penetration into the base stock or high evaporation rates involves complex changes in flow, outlined in Chapter 3.

The type of flow describes the effect of increasing force on the fluidity of the coating. A plot of the fluidity with increasing force, followed immediately by a similar plot with decreasing force, forms two curves. The loop or area between the curves is a measure of the thixotropy of the coating. This loop can be interpreted to mean that the structure of the coating is broken with increasing force and that reestablishment of the structure does not occur promptly under decreasing force. This failure to reestablish the structure immediately is responsible for the loop that is typical of thixotropic behavior.

Most aqueous coatings approach Newtonian flow at low solids, a condition that is predominant in coatings metered by an air-knife coater. These coatings usually contain synthetic rubber latex dispersions stabilized with surface-active materials that tend to produce low viscosity and good leveling properties.

Migration of Components

Although ingredients are uniformly distributed throughout a fluid aqueous coating, redistribution or migration of the components can occur upon application of a porous absorbent substrate to the surface. A high solids layer or filter cake of pigment particles is believed to form near the interface between the paper base and the coating layer. Particles of pigment or adhesive with diameters of $0.1-1.0\,\mu$ can penetrate the base paper with pore sizes at $0.45-0.8\,\mu$.

High-solid coatings are less prone to migration into the base paper and therefore remain more completely on the surface. The internal sizing, wetability, and pore size of the base paper influence the tendency toward saturation. Conditions that promote migration of adhesive into the base paper result in a greater use of adhesive for sufficient pigment bonding. Starch adhesives penetrate into the base stock more than casein and other proteins on any given base paper.

The depth of coating penetration is also influenced by coating properties such as surface tension and viscosity. These paper and coating factors are mathematically related as follows:

$$l^2 = \frac{\lambda r t \cos \theta}{\eta}$$

where l is the depth of penetration, λ is the surface tension of the coating, r the radius of the paper pore, t the time of contact, θ the contact angle or wetability, and η the viscosity of the coating fluid. In a practical sense, coating holdout, which is necessary for a good printing surface, is attainable by regulation of the paper and coating variables shown in this equation.

Figures 4-1 and 4-2 illustrate the relationship between (a) temperature and water retention in the coating and (b) percentage of solids and water retention.

Recently, progress toward improved surfaces has been obtained by double- and triple-coated papers. Each coating can be properly formulated to impart special characteristics to the sheet. The prime coating may be designed to improve hold-out, increase opacity, level the surface, and improve interfacial bonding. The ingredients in different strata can be selected to maximize surface perfection and minimize cost. Multiple coating layers make it possible to avoid excessive penetration and obtain maximum efficiency of adhesive in the top layer. Reduction of adhesive in the top stratum improves finishing and ink absorption.

Effect of Drying

The distribution of coating materials in the layer is also influenced by the rate of water removal during the drying cycle. High evaporation rates tend to direct the soluble or colloidally dispersed substances toward the surface. A coating at low solids and high drying rate may be rich in solubles at the surface, whereas the same coating dried slowly may promote penetration into the paper. Soluble adhesives, when concentrated at the surface, can bar ready access to the pigment; when concentrated at the base, they bond the pigment insufficiently in the coating. Concentration of the pigment at the base may lead to insufficient bond between the coating and the paper. Most researchers believe that adhesives and solubles follow the water flow within the coating layer before the coating loses enough water to become immobilized.

Paper coatings composed of mineral pigments or other dispersed materials, plus casein and latex at 36—50% solids can suffer serious damage when dried in excess of 2 pounds of water per square foot of drier per hour. The damage manifests itself in three ways: First, the rate of ink absorption becomes low and nonuniform; Second, resistance to picking of the coating during printing decreases; Third, the coating cracks, acquiring the appearance of "crazed" ware as seen in old china. The three effects are cumulative—by the time the second appears, the first is already present. The third is always accompanied by the first two. The net effect on printing and special functional papers is unacceptable quality.

At any given evaporation rate, damage is less likely to occur as the coating solids increase and the coating weight decreases. The damaging migration occurs during a portion of the constant-drying-rate period while the coating is still mobile. It may begin at the

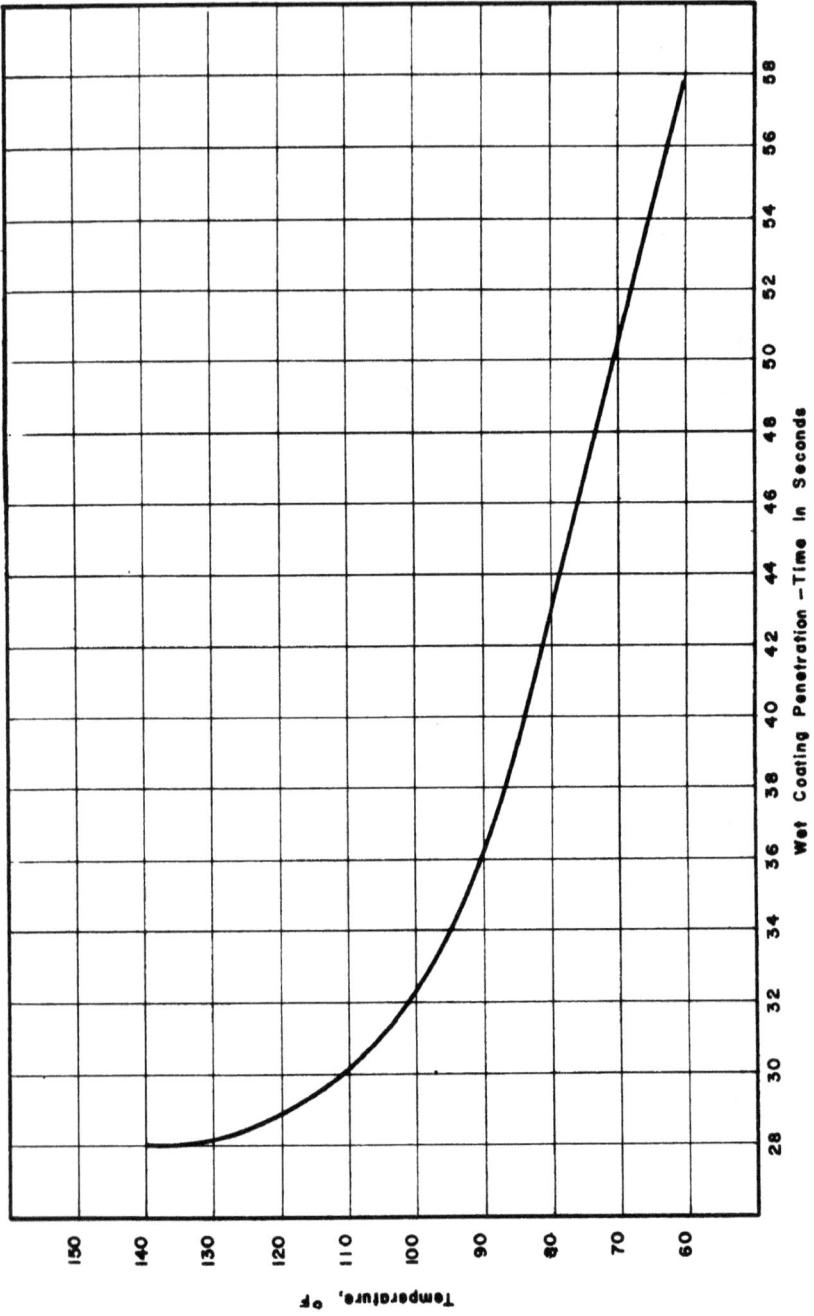

Fig. 4-1. Temperature *vs.* penetration time [36% solids; Argentine casein pigment; 60-1b book paper base].

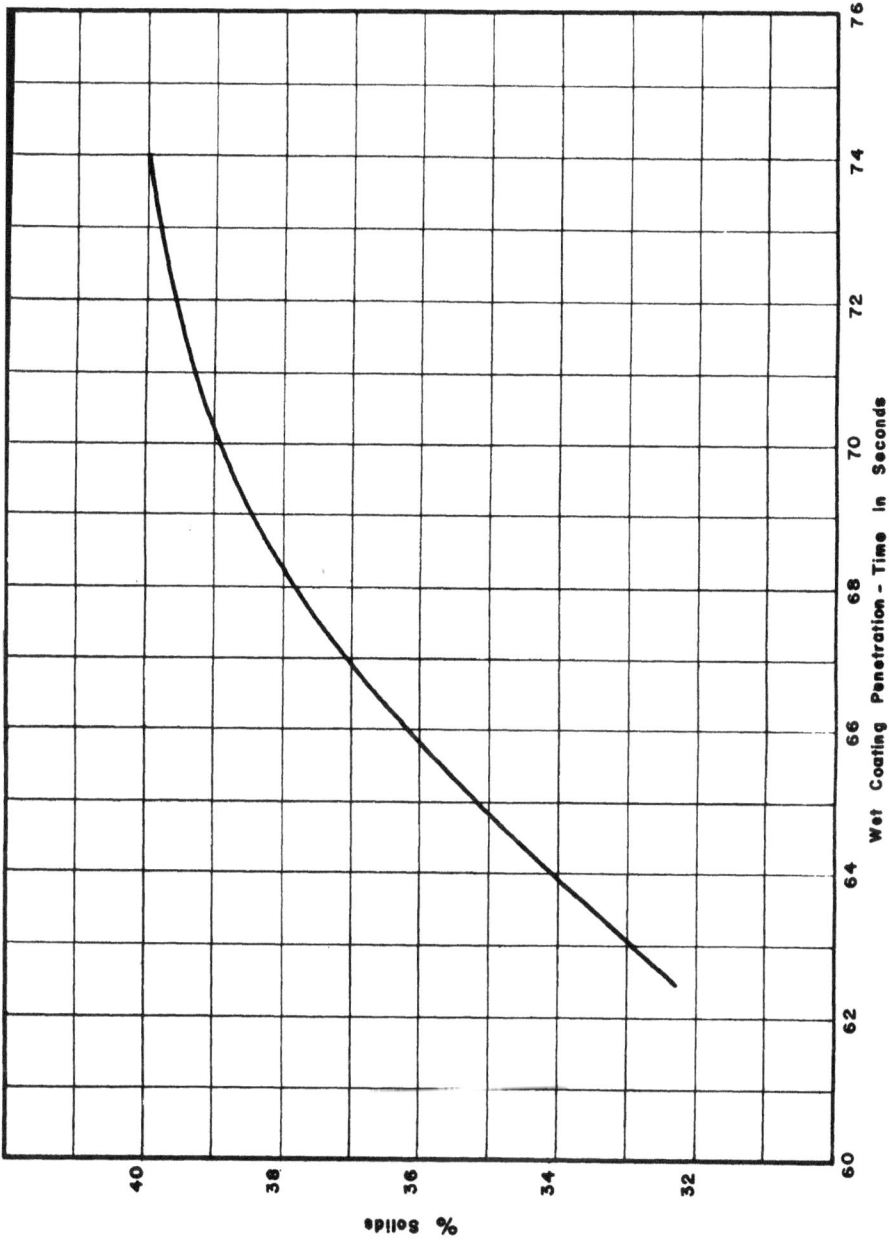

Fig. 4-2. Percentage solids *vs.* penetration time at 74°F. [Argentine casein pigment; 60-1b book paper base].

point when paper plus coating arrive at the temperature that the drying mass retains during the constant-rate period. It may end in the early stages of the falling-rate period. As the web proceeds well into the falling-rate period its temperature begins to rise. The topmost surface layer then approaches the temperature of the drying air. A large differential in both temperature and moisture content throughout the coating may be the cause of cracking.

Mass and Volume Effects

After calendering, properties of the finished coated paper, such as adhesion, brightness, opacity, and gloss, depend to some extent on the physical weight and volume relatiohship of pigment and adhesive. The quantity, size, and shape of the pigment have a dominant effect on the quality of the surface. The relatively small quantity of adhesive may not be adequate to occupy completely the voids between particles and also to bond the layer to the base stock. Expressed as a percentage, *Pigment Volume Concentration* $= \dfrac{100\ p}{(p+a)}$, where p is the volume of the pigment, a is the volume of solids in the adhesive, and $p+a$ is the total volume of solids. High percentages of pigment volume concentration indicate high pigment and low adhesive content. The level of pigmentation at which coating adhesive solids just fill all voids between pigment particles is called *Critical Pigment Volume Concentration*. Willets and Marchetti obtained the following results for several pigments by the oil-absorption method:

Critical Pigment Volume Concentration, %

Fine particle size clay	43
Coarse particle size clay	50
Fine particle size precipitated $CaCO_3$	50
Water-ground calcium carbonate	69
Titanium dioxide (rutile)	51

The pigment portion of most printing paper coatings is usually in excess of 65% of the total volume of solids, well above the critical pigment volume concentration. This value indicates that the adhesive solids are not of adequate volume of fill the space between pigment particles in the dry coating. Brightness and gloss after calendering rise with increasing pigment volume concentration, which shows the dominance of the pigment.

The relative void volume in a particular coating is not significantly affected as the size of the pigment particle is increased. Rather, the size of the individual voids increases and the number of voids decreases. Numerous tiny voids exist when the pigment is finely dispersed. Under these circumstances the adhesive may be insufficient to be uniformly distributed along the surfaces of all voids. Consequently, some portions are unbounded and the pick resistance is low. Reducing the number of voids and increasing their size by using larger pigment particles may lead to improved bonding.

The concept of relative sediment volume, which measures the quality of pigment dispersions, has been introduced by Robinson. The volume occupied by the pigment in a water-base dispersion when the particles have settled against each other by gravity or centrifugal force is known as sediment volume. It is equal to the sum of the volumes of the solids and the voids. Coarse particles produce large sediment volume, whereas fine particles yield small sediment volume. The sediment volume divided by the volume of solid pigments contained in the total sediment volume is known as the relative sediment volume. Reduction of the relative sediment volume to the lowest possible level is equivalent to effective dispersion.

These concepts have served well to destroy the myth that the adhesive requirement of a pigment is determined by the specific surface of the pigment particles. The percentage of voids in a given volume of pigment is mainly responsible for the adhesive demand. Most pigments are adequately bonded, one to another, when 20—25 % of the volume of the pigment voids is occupied by adhesive.

Certain practical considerations such as basic properties of the coating solids, method of application, and rate of drying may cause increases in adhesive requirements beyond the theoretical pigment–binder ratio.

Emulsions

Water-base coating systems sometimes exist entirely, or partially, as emulsions. When two immiscible liquids are combined, one of the liquids takes on the form of droplets dispersed in the other liquid. Such a system of fluids is known as an *emulsion*. Common practice describes emulsions according to the way in which the aqueous phase is distributed. When water constitutes the continuous phase, the emulsion is called an oil-in-water emulsion. The term oil applies to any water-immiscible liquid phase, regardless of its characteristics. When the water phase is the internal or dispersed phase, the system is called a water-in-oil emulsion. More complex emulsions may

also exist in which oil is dispersed in an aqueous phase and each of
the oil droplets may contain a number of tiny water droplets. Two
immiscible organic liquids can also form emulsions.

Stable emulsions are produced by a large variety of different types
of water-soluble and oil-soluble soaps and synthetic nonionic, anionic,
and cationic surfactants. Naturally-occurring emulsifying agents
such as phospholipides, gums, and proteins are also used. Finely
divided solids partially wetted by both phases are also effective. Traces
of electrolytes stabilize extremely fine dispersions of small quantities
of oil in water, known as hydrosols.

Particle size and size distribution and optical and electrical properties of emulsions determine their characteristics and usefulness
for intended applications. The lower size limit of emulsion droplets
is placed at the smallest size that can be seen with an ordinary optical
microscope—about 0.1 μ; the largest size of an emulsified droplet is
set at the limit of visibility with the unaided eye—about 50 μ. Most
commercially important emulsions have a particle size range of 0.1
—10 μ.

The appearance of an emulsion is determined by the particle size
of the dispersed droplet and the difference in the refractive indices
of the two phases. The emulsion is clear or transparent when each
phase has the same refractive index regardless of particle size. When
the refractive indices are different, the opacity of the emulsion increases with a decrease in particle size, the maximum opacity occurring when the particle diameter is 1 μ.

Microencapsulation

Emulsion technology has recently advanced to a state of refine-
ment where indivdual microscopic droplets can be surrounded with
a thin layer or covering. A chemical or mechanical process known
as microencapsulation, is one in which liquids or solids are encased
with a membrane or film structure in the form of microscopic capsules.
Although somewhat arbitrary, microcapsules most commonly exist
in sizes up to 1/25 in. in diameter. Various techniques for capsule
production exist; these may produce capsules ranging from a few
microns to half an inch or more in diameter. Such a process makes
it possible to enclose in microscopic spheres materials that are to be
protected and subsequently released at a chosen time under specific
circumstances. The features of size and controlled release make the
concept and process useful in paper coating and other applications.

A popular and significant application in coated paper is the carbonless copy papers now available.

Chemical microencapsulation requires that a film-forming material, dissolved in the continuous water phase of an emulsion, be deposited on the surface of each insoluble droplet of the dispersed phase. Coacervation is the process of forcing dissolved macromolecules (large polymeric molecules) out of solution and to change from a liquid phase to a dispersed coarse mixture. A salt or some other suitable foreign component is added to the solution to reduce the solubility of the polymeric material and force it out of solution. This insoluble liquid substance, generally gelatin, forms the walls of microcapsules produced in this manner.

Coacervation can also occur when two different kinds of macromolecules unite to form a colloidally dispersed insoluble polymeric substance.

Whether simple or complex, coacervation produces an insoluble liquid substance that has the freedom to assume the shape of the material to be encapsulated. Polymeric materials such as gelatin, when cooled sufficiently, form a solid gel. This phenomenon is utilized in causing the liquid gelatin to be converted to a solid wall.

Conveniently, colloidal solutions of gelatin macromolecules are good emulsifying agents. A solution of colloidal gelatin is soluble in the water phase and only slightly soluble in the oil to be emulsified. Although the gelatin molecules stay in solution in the water, they are also attracted to the oil so that they form a protective layer around each oil droplet, thus forming separate capsules.

The forming of emulsions is functional, for when the time comes to make coacervation take place, the gelatin molecules are already situated around the oil droplet. When the gelatin molecules become insoluble, they naturally tend to take the shape of the emulsified oil droplets, surrounding each globule with a wall of more-or-less porous gelled colloidal material.

The first step in the process involves dispersion of the chosen hydrophylic colloid in water. To this dispersion, the oil that becomes the nucleus of the capsule is added. The mixture is agitated to form a colloid–oil–water emulsion. The coacervating salt is added to the emulsion in the appropriate quantity. The salt renders the colloid less soluble in water, causing a fluid membrane of colloid to surround the oil droplet. During these steps, the temperature is carefully regulated above the melting point of the colloid material. The capsule wall is solidified by cooling to a point where the colloidal material solidifies. Washing, filtering, and hardening of the capsule walls with formaldehyde completes the process. The capsules can be re-immersed in water as an emulsion, or spray-dried in the form of a freely-flowing powder.

Bubble Coating

A novel emulsion process has been commercially perfected to provide a composition of extremely high brightness and opacity. The coating layer is composed of an adhesive matrix containing numerous air bubbles of uniform size, homogeneously distributed throughout the film. For a white or colored surface one can produce an opaque coating composition that does not require the use of pigments to achieve a high degree of light scattering.

An emulsion of water and oil in combination with an alkali-soluble protein forms the liquid coating system. The film-forming matrix material in water is the continuous phase and a water-insoluble liquid is the discontinuous phase. The immiscible liquid must have a boiling point above that of water. Modifying agents, used to insolubilize the adhesive, may also be dissolved in the liquid phase so that the adhesive is converted to an essentially water-insoluble condition in the final coating.

Composition

A typical composition for bubble coating is formed by preparing a water suspension or solution of a proteinaceous adhesive such as casein. A suitable water-immiscible liquid such as kerosene, mineral spirits, or a high-boiling acetate, is then added under conditions that will provide the desired emulsion. The globules forming the discontinuous phase may range in size from about 0.5 to 1.5 μ.

The liquid chosen to form the discontinuous phase may also require consideration of factors such as color, compatibility, toxicity, flammability, cost, and adaptability to production procedures. In some instances, small quantities of other adhesives or particulate materials such as inorganic pigment may also be added. Although generally formed entirely of proteinaceous material, elastomeric polymers are also mixed with the natural adhesive.

Application

The liquid coating is applied by conventional coating methods as an emulsion. During drying, a portion of the water is removed, thereby gelling the adhesive matrix, which contains numerous small droplets of the discontinuous phase uniformly distributed throughout the layer. Finally, the remaining water and organic liquid are removed through additional heating. The remaining coating consists of an adhesive matrix with numerous voids profusely distributed throughout, the largest bubble preferably not over 5 μ. The multitudinous minute air bubbles produce a uniformly porous structure, providing air-adhesive interfaces that contribute intense brightness and opacity to the layer by virtue of their light-scattering ability. At the same time, the overall weight of the coating is reduced by virtue of the substitution of air for high-density pigments normally used in paper coating.

Bubble coatings provide a composition suitable for high-speed application to a paper web to form an opaque surface with good printing properties. The system is particularly well suited to publication papers, inasmuch as good optical properties, light-weight coated papers, and high-speed production can be achieved.

Polymer Dispersions

Another fluid system commonly encountered in the paper coating industry is synthetic and natural high-polymer, film-forming, water-base emulsions and dispersions. A synthetic or natural insoluble polymer can be carried in a water-base system as a colloidal dispersion or an emulsion.

In either case, we deal with a two-phase system wherein the continuous phase is water and the dispersed phase has film-forming properties similar to those formerly available only through solvent coatings. Water-insoluble resin films, clear or pigmented, can be produced from a variety of materials to obtain a wide range of functional properties, without the disadvantages of health and fire hazards or toxicity problems common to solvent coating. Liquid and vapor barriers, air impermeability, scuff-and-abrasion resistant, acid-and-alkali safe, and weatherproof coating layers are attainable from such aqueous base systems.

Emulsions and dispersions are available at a solids content of 40—65% at suitably low viscosity, thus reducing the drying requirements in time and machinery. High-molecular-weight, strong, and flexible films are available without a compromise between film properties and operating conditions, as is the case of solvent coatings. In a water dispersion, the particle size, dispersion methods, and particle packing determine the solids content and viscosity limits of the coating.

Recent advances in colloid technology have overcome earlier problems associated with physical and chemical causes of instability of emulsions and dispersions. Tolerance to freezing, mechanical shear, pH variation, and additions of polyvalent metal ions have broadened their useful applications.

Emulsions

Formation of continuous films from dispersed systems usually requires a high-temperature exposure following drying to fuse the discrete particles. This procedure may present operating problems with over-dried brittle paper, curling, and frequent breaks during the coating operation. Another difficulty with such colloidal systems is foam, which leads to pin holes in the dried film. The emulsifying and stabilizing surface-active chemicals, which are incorporated to obtain stability, are responsible for this difficulty. These additives also contribute to the hydrophilic nature of the coating film, thereby making water-insensitive films difficult to produce. The polymer suspensions are also irreversible, forming solid phases that cannot be redispersed once they have been coagulated.

Dispersions

Dispersions of synthetic resin latex are formed by emulsifying the solid components or by polymerizing the chosen monomers, either alone or in combination. All commercial methods for preparing resin dispersions involve the reduction of the solid material into discrete colloidal particles. A surface-active material or protective colloid is required to maintain a stable suspension of individual particles. Dispersions or emulsions of nitrocellulose, ether–cellulose derivatives, natural resins, rubber derivatives, alkyd resins, and various polyvinyl derivatives can be produced in aqueous base systems.

The choice of kind and form of coating materials or systems is the result of an attempt to satisfy many requirements besides the eventual application. The materials selected determine the coating-color behavior at each unit process in the manufacturing operation. Details such as the nature, size, shape, and concentration of dispersed particles, whether pigment or film formers, and their interrelationship with other components influence the operating characteristics. Consideration is given to the kind of mixing, coating, drying, and finishing machinery available. In specialty coating, the application machinery, whether it is roll, air-knife, or trailing-blade plays an important role in establishing certain flow- and solids-content limits for optimum operation. It is desirable to strive for the highest possible solids content to reduce the drying load and minimize coating penetration into the base paper. The rawstock to be processed places certain requirements on the formulation to cope with variations in formation, surface contour, internal sizing, physical strength, and optical properties. The variations that can be tolerated by the formulation must be known and understood, for total correction of deficiencies on major base stock is not attainable. Economic factors related to the market requirements place a definite limit on the quality of pigments and adhesives available to the formulator.

BASE PAPER

The base paper used for water-base coatings can often be the most important and troublesome element to the coater. Regardless of kind or grade, paper represents the major portion, by mass and volume, of the coated sheet. In view of the trend toward lighter coating weights, the dominance of the base paper is becoming increasingly greater. In some instances, the base paper is made in the same plant that uses it, but independent coaters purchase it from paper mills. A conversion or specialty coater is likely to have more direct control over the coating and machine characteristics than over the base paper. The independent coater therefore attempts to set specifications for the paper maker that accurately describe the requirements for a rawstock that performs well on the coater and meets the many eventual requirements. Unfortunately, the specifications generally involve tests that only tell the paper maker something about the *paper* he is making, and do not accurately measure or predict the performance on the coater. Many of the properties required for optimum performance are of an intangible nature.

A specialty mill is likely to coat papers that range in basis weight from a 25-pound gift wrap to a 20-mil tagboard or cover. The fiber furnish may include sulfite, soda, and kraft wood fibers, as well as rope or manila hemp for unusual strength. Nevertheless, certain requirements are necessary and similar for all papers. The present trend is toward precoating on the paper machine with size press, gate roll, or trailing blade with one, and sometimes two, layers on one or both sides to provide a smooth and economical undercoat ahead of conversion coating. This practice has contributed toward better control of the surface smoothness and controlled hold out. It allows for concentration of premium ingredients in the top surface only, thus providing opportunities for economic utilization of coating materials.

General Requirements

The following important general properties have evolved from the point of view that emphasis should be directed toward producing a base stock to fit the coating, rather than the converse:

1. The paper should possess the ability to remain flat while passing through the coating machine. Excessive curl, cockle, or physical distortion during coating causes coating streaks and edge cracks that lead to breaks and inefficient operation during coating and calendering.

2. Coated one-side paper as used for labels should lie flat during coating and should not curl or change dimensions when subjected to changes in moisture content.

3. The base paper should accept the coating and yield a level, continuous surface that, upon calendering provides a surface of uniform gloss and acceptable printability.

4. An acceptable coating bond between the coating layer and rawstock surface should result without excessive use of adhesive.

5. Optical properties such as brightness and opacity should be adequate to meet final requirements of the finished product without excessive coating weight.

6. The base paper must have sufficient dry and wet strength to (a) withstand stresses during coating and in the long draw of a two-side coater, (b) resist splitting when printed by offset lithography, and (c) survive the rigors of high-speed printing operations.

Despite a difference of opinion on the significance of various tests, most investigators agree that uniformity of sizing, density, and formation are desirable qualities. Machine- and cross-direction characteristics and the character of felt and wire sides should be the same.

The density of the sheet appears to be one measurable characteristic that provides some basis for predicting performance of the paper on the coater—the least dense performing the best.

Uniform internal sizing of the base paper is extremely important in order to minimize the coating penetration and reduce the amount of coating required to obtain continuity of film. Absorbent paper causes excessive drainage of adhesive, which in turn leads to weak coating bond.   There is a minimum internal sizing, below which surface defects occur in the coating.   The photomicrographs in Figure 4-3 illustrate the effect of low and satisfactory sizing on the coated surface.

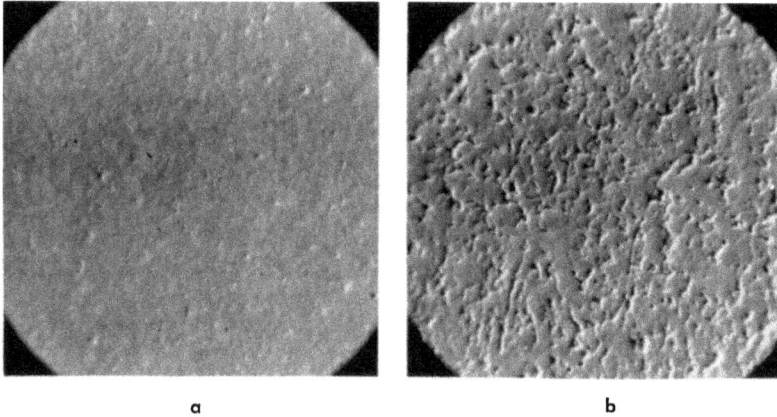

a   b

Fig. 4-3.   Extreme case of pebbly coating.   K&H galvanic sizing: (a) 51 sec; (b) 13 sec.

The best possible and most uniform formation consistent with the strength requirements and the nature of the fibers is mandatory. Nonuniform wetting and absorption of the coating occurs when the formation is wild or bunchy, which leads to a disagreeable mottled appearance when the paper is supercalendered.   Gross amplification of this irregularity is apparent on colored coated papers.   Poorly formed papers cause small-scale variations in density in the thick and thin areas when the paper is supercalendered to a uniform caliper. This situation causes nonuniform absorption of printing inks that manifests itself in mottled appearnce in solid black or colored printed areas.   The most uniform formation produces paper with low fold and tear.   Therefore, a compromise between strength and good formation usually results.

Physical Properties

Surface smoothness has come to be more significant that finish. Precoating techniques with adhesive films or pigment coatings provide a plane surface that results in level top coatings of uniform thickness.

The preliminary coating layer serves well to eliminate surface fuzz and to establish a surface with uniform wetability and absorption for subsequent coatings.

Uniformity of caliper is essential to insure uniform coating application and satisfactory performance on the coater and calender. Lower-than-average caliper along one edge usually causes a loose or slack edge that makes it difficult to maintain uniform tension across the web in the coater. Those areas at low tension usually accept less coating, and in some instances lead to skipped portions. Such defects occur at all subsequent unit processes and result in excessive waste because of cuts or breaks in those areas. Variations in thickness also cause the coated paper to take a different gloss on supercalendering and to print differently on the press.

Color, opacity, and brightness are usually specified to control the base stock as near the finished paper as possible. A great contrast between the base paper and the coating favors a mottled appearance. This contrast is particularly important when light coating weights or slight variations in coat weight are likely to occur.

Good mechanical condition of the roll regarding uniform hardness, freedom from calender cuts or creases, flawless slitting, and freedom from minor edge cracks is necessary to assure continuous operation at the coater. Defects, mechanical in nature, lead to breaks in the web; these cause excessive waste and considerable lost production time. Holes, slime spots, shives, and excessive dirt also produce faults in the finished paper.

Typical specifications that attempt to describe the desirable characteristics for coating base stock are shown in Table 4-1.

COATING INGREDIENTS

Detailed descriptions of coating ingredients follow.

Water

The conventional fluid-coating process requires a polyphase wet-coating system. Water is the liquid phase, which acts as a carrier for the dispersed or dissolved components. The dispersed phase consists of an adhesive and pigment mixture. Additives are often used to control operating characteristics and to enhance the quality of the final paper.

Each classification of materials includes a large variety of compounds, all of which contribute in some way toward the total coating process. Water is a major ingredient in the low-solids coating systems commonly used in conversion coating for specialty grades. Certain of the difficulties encountered because of incompatibilities in complex colloidal aqueous coatings can be traced to the quality or variations in the quality of water. A small addition of a polyphosphate or another sequestering agent to hard water may prevent the formation of water-insoluble compounds formed from calcium and acidic ions present in adhesives. This mechanism can prevent secondary thickening of a coating solution.

The chemistry of the local water supply plays an important role in the preparation of pigment slurry and adhesive solutions. Seasonal variations in the type of water treatment may influence the dispersing agent and alkali requirements. Extremely hard water may inactivate the dispersing agents that are sometimes included with predispersed or water-dispersible pigments. Savings sometimes can be realized by using soft or demineralized water. Depending on the local water source, the treatment may include additions of copper or other metallic salts to control growth of algae during the summer season. These polyvalent metal salts, even in trace quantities, sometimes interfere or create undesirable color reactions in special coatings.

Adhesives

The suspended or dissolved solids of a coating layer are bonded to a fibrous bodystock by an adhesive. Such materials, carried in water, provide a liquid vehicle to suspend the pigment during the coating process, to fill the voids between particles of pigment, and to bind them one to another and firmly to the paper. Many theories attempt to describe the complex phenomenon of adhesion between two solid surfaces; the strength of the bond may be influenced by the kind of adhesive, wetability, perfection of contact, concentration of impurities, and the chemical environment.

Adhesives also influence the rheology and water retention properties of a coating mixture. The amount and distribution throughout the layer affects the absorption of the ink vehicle during printing and the varnish penetration in overprinting.

Although adhesives represent a relatively small proportion of the coating by weight and volume, their optical properties, such as brightness and refractive index, do contribute to the properties of the final coated paper.

Chemically, the natural adhesives are carbohydrates, in the form of starch or modified cellulose, and proteins derived from casein, soybeans, and animal tissue. Synthetic adhesives such as styrene–butadiene,

Table 4-1

Desirable Characteristics for Coating Base Stock

Test	Bond Base	Book Paper	Cover Paper	Test Method
Basis weight, 25×38-500	25 lb±1.3	60 lb±3.0	110 lb±5.5	TAPPI, T410m-45
Thickness, 1/1000 in,	2.2±0.2	4.4±3	7.3±0.4	TAPPI, T411m-44
K & H sizing, seconds	Av. 28-40	Av. 65-90	———	Galvanic Test
Air resistance, sec/(100 ml)/(in^2)	Av. 20-35	Av. 30-35	———	TAPPI, T460m-49
Brightness	78 min.	84 min.	78 min.	TAPPI, T462m-58
Opacity	64 min.	90 min.	———	TAPPI, T425m-60
Internal tear, g/sheet, M.D.	20 min.	50 min.	180 min.	TAPPI, T414 Ts-64
MIT fold, C.D.	———	40 min.	150 min.	TAPPI, 423m-50
Surface strength (wax)	12 min.	14 min.	14 min.	TAPPI, T459m-48
Smoothness	200 max.	140 max.	120 max.	Sheffield
Water absorbtiveness (Cobb), g/m^2	15-25	15-25	15-25	TAPPI, 441-05-63

acrylics, butadiene–acrylonitrile, and vinyl acetate are available as dispersions or emulsions in water. Water and alkali-soluble synthetics such as polyvinyl alcohol and polyvinyl acetate are offered in dry form.

Natural adhesives are used in large volume–approximately 200,000 tons per year for pigment coatings. The use of synthetic products is growing rapidly, having taken over about 10% of the total. The synthetic products are generally two to five times as expensive as the natural products. Therefore, on a dollar basis, the synthetic products represent about 30% of the total dollar volume expended for coating adhesives.

The major property to be considered in selection of an adhesive is its bonding strength. The amount required to bond a given pigment determines the final properties and economics of the coating formulation. Practically speaking, bonding ability is purchased when a particular adhesive is chosen. In addition, low specific gravity, high brightness, and compatibility with other adhesives is essential. Ease of solubility, preferably in cold water without addition of alkali, stability of viscosity, and dispersing power are desirable attributes. An adhesive free from decomposition by bacteria and fungi, low-foaming, free from odor, and easily insolubilized contributes greatly to smooth manufacturing operations.

Animal Glue

Animal glue is an organic colloid derived from collagen—a protein constituent of hide materials and bones. Animal glue, despite its long history, did not realize its maximum utilization until progress in protein and colloid chemistry established the substance in select areas of the coated paper industry. An animal glue film, deposited from a water solution, is continuous, noncrystalline, strong, and elastic. Its tensile strength is calculated to be in the range of 10,000 lb/in.^{2}, denoting its potential to make tenacious bonds. Because of its peculiar protein structure, animal glue is not precipitated by acids or alkalis and also resists structural breakdown by such materials. The film is soluble only in water and especially resistant to grease, oil, alcohol, and other nonaqueous fluids. It is without equal as a protective colloid and can be made extremely moisture resistant by treatment with tanning and insolubilizing agents. Such treatment is commonly done in making blueprint, chart, and currency paper, and washable wall papers. The usual tanning compounds are formalin and formaldehyde donors such as paraformaldehyde, hexamethylenetetramine, chrome alum, and mixtures of aluminum sulfate, sodium acetate, and borax, which may be applied as an overwash, or incorporated directly in the formulation of the glue. On blueprint paper, the insolubilized glue surface protects the paper against wet disintegration, yet provides a surface film adequately absorbent to the chemicals used in blueprinting. On chart paper, this feature prevents "bleeding" of ink, protects the paper, and absorbs the ink without "feathering."

The value of animal glue to the paper coating industry, past and present, is derived from its protective colloid action, ease of preparation and application, versatility of use, ability to form strong and elastic films, colloidal reactivity under a wide range of alkaline and acidic conditions, and its permanence of adhesive bond.

Animal glue was the first, and for many years the only, adhesive used for bonding pigments in coated papers. Subsequent development brought other natural and synthetic adhesives into the field. In recent years, animal glue has been successfully applied in particular specialty coating applications, including colored and metallic pigmented papers. Among the important characteristics in these applications is the amphoteric nature of the colloid, which makes it compatible with pigments, dyes, and base stocks in acid, neutral, or slightly alkaline systems. Upon drying, the coating layer is noncracking, an important requirement for colored coated papers.

The trend in the coating industry is to use water or steam-jacketed monel or stainless steel cookers for the preparation of glue solutions. Variable-speed agitators and thermostatic temperature controls are preferred. The solution is usually prepared at 12% concentration, one pound of glue per gallon of cold water for soaking and swelling. Melting occurs at 140°—145°F, modifying chemicals are added, the solution is diluted to the desired concentration, and controled at 105° —120°F during application.

Casein

Casein, a natural polypeptide polymer, is precipitated from skim milk by the action of natural souring, acids, or rennet. At the isoelectric point of casein, near a pH of 4.5, the casein is least soluble and coagulates readily. Lactic acid—which can be generated from the sugar in milk—hydrochloric acid, and sulphuric acid are commonly used to precipitate the protein.

Even at current high prices, casein is extensively used in off-the-machine conversion coating because it excels in brightness, ease of waterproofing, durability, printability, and operating characteristics.

Casein is often blended with starch or latex emulsions, especially where high solids and good flow are required.

Chemical properties. The most important chemical properties of casein to the paper coater are the carboxyl acid-reactive and the amino basic-reactive natures of the molecules. With alkalis, this reactivity makes it possible to form soluble salts that can be subsequently replaced with polyvalent calcium, aluminum, or zinc ions to produce an insoluble film. The casein molecule is also reactive with organic compounds such as formaldehyde, aldehyde donors, and acid anhydrides. Such reactions block the cationic groups or cause cross-linking between molecules to produce insolubility in water. Casein, because of its amphoteric nature, is soluble in acids or alkalis, although coating mills generally use alkali to obtain solution. At low cost, sodium hydroxide produces solutions at low viscosity with minimum alkali; good bonding ability and films with high strength are obtained. With such a strong alkali, pH control is difficult, water sensitive films result, and bacterial decomposition is rapid. Excessive use of sodium hydroxide causes darkening and hydrolysis of the protein molecule, with weakening of the casein bond. Residual alkali also attacks the rawstock, causes colored pigments to change hue, and tends to produce scumming in the lithographic printing process.

Solubilizing Casein. Ammonium hydroxide is frequently used to solubilize casein, with only a relatively small quantity needed for complete solution. Its cost is substantially higher than that of sodium hydroxide. Being volatile, the ammonia is largely driven off during the drying cycle, leaving a film that is more water resistant than when sodium alkalis are used. Good pH control and low viscosities are attainable by taking advantage of the buffering power of ammonia. Casein, solubilized with ammonia, is more tolerant at a pH between 7.0 and 8.0 to formaldehyde additions without excessive thickening. Frequently, ammonia and borax are used in combination in order to benefit from the advantages of ammonia and the preservative power of the borax.

Sodium carbonate is quick acting, yields a strong film, has excellent buffering ability, and is cheap. Possible evolution of carbon dioxide increases the tendency to foam, and causes high viscosity and poor resistance of the casein film to water.

Alkalis such as trisodium phosphate, sodium metasilicate, hydrated lime, and sodium fluoride are occasionally used to solubilize casein. Each system has advantages and disadvantages that must be considered and made a matter of compromise in making a selection. Whatever

the choice, the total amount of alkali used is determined by the quantity
necessary to dissolve the casein, and the desired pH level of the coating.

Urea and dicyandiamide are used quite generally to thin or liquefy
casein solutions to provide workable viscosities at high concentrations.
These materials, used at 10—20% of the casein, reduce the initial viscosity, and permit solutions that contain up to 33% solids to be prepared.

Methods of preparing the casein solution vary, but usually involve
stainless steel vessels equipped with agitators, a steam supply (direct
or indirect), and a means of cooling. The casein granules are added
to cold water with agitation to prevent large agglomerates from forming.
Approximately 15 minutes are allowed to wet thoroughly and swell the
casein particles. The alkali is added and the system heated to approximately 55°C and held until solubilization is complete. The solution
is then cooled quickly to minimize darkening and loss in adhesive strength. The solids in solution usually amount to 15—20%; the solution
has a final pH between 6.5 and 9.5.

In recent years, skim milk has gained a strong position as an important food product, thus emerging from its earlier role as a by-product
of the dairy industry. This development has created a competitive
situation that has diverted the skim milk to food production and forced
the price of casein to unprecedented high levels. As a result, paper
coaters are vigorously exploring natural and synthetic substitutes for
casein.

Soybean Protein

Soybeans contain a crude protein content near 40%. Commercially,
the beans are cracked, dehulled, flaked, and extracted with hexane to
remove the oil. Soybean protein is alkali-soluble and acid-precipitable,
showing a chemical and physical similarity to milk casein. In many
paper coating applications, casein and soybean protein are used inter-
changeably, with only minor alterations in the formula and the process.
Casein is claimed to have better color, greater ease of waterproofing,
less tendency to foam, and a stronger bond than soybean protein. The
latter, being produced in large-scale plants, is more uniform in quality,
allows for more flexibility in use because different viscosity grades are
available, and is less subject to market fluctuations than casein.

Like casein, the soybean protein is soluble in aqueous alkalis such
as sodium hydroxide, sodium carbonate, ammonia, and borax. Preparation of solutions, quantities of alkali, insolubilization, viscosity re-

duction, and preservative action are similar for soybean protein and casein.

Soybean protein of interest to the paper coater is available in three forms: (a) unhydrolyzed or unmodified types; (b) hydrolyzed grades of high, medium, low, and extra-low viscosity; and (c) chemically modified grades. The unmodified form is obtained by extracting the protein from the flakes at mild temperature and pH conditions to minimize any tendency toward hydrolysis. The unhydrolyzed form is dispersed with alkali and held for several hours at high temperature and pH, during which time the protein undergoes extensive changes in dispersibility and viscosity. This action, depending on the variations in pH, time, and temperature, leads to various degrees of hydrolysis. Proteins with a range of stable viscosities and desirable flow properties similar to casein are produced by this technique.

A soybean protein with improved bonding ability and lower viscosity is obtained by acylation of the hydrolyzed protein. During the hydrolysis action, the acylation is accomplished by adding monocarboxylic or dicarboxylic acid anhydrides to the alkaline dispersion of protein. Other forms of chemically modified proteins have been obtained by reaction with organic compounds. Reaction with sodium chloroacetate yields a carboxymethylated protein that is resistant to bacteria and does not gel on addition of formaldehyde.

Alterations and modifications to the soybean protein have made stable adhesives of low viscosity available for high-solids coating preparations that perform well on roll and trailing blade coaters. These recent developments permit the paper and paperboard coaters to use large quantities of soy protein with satisfactory results.

Starch and Starch Derivatives

Starch for paper coatings is principally derived from corn, although large amounts from potatoes, wheat tapioca, and waxy maize are also used. Chemically, the starch molecule is a natural polymer of glucose units arranged in pyranose rings held together through an α-glucosidic linkage. The glucose units form polymer chains in linear and branched configuration. The straight-chain construction is known as the amylose fraction, whereas the branched structure is termed the amylopectin fraction. The distinction is primarily a physical one, the chemical properties being related. The branched molecules are larger and have much higher molecular weight. Corn starch usually contains approximately 75% amylopectin and 25% amylose. Hybrid varieties have been developed to control the proportion of linear and branched fractions. Some waxy maize varieties of corn contain only amylopectin and no amylose.

Configuration plays an important role in the stability of the starch solution. Unmodified corn starch, dispersed in water, tends to gel and to assume the original undispersed state upon ageing. This phenomenon is described as retrogradation and is a troublesome property to the paper coater because it causes undesirable changes in viscosity and partial sedimentation of components. The straight chain amylose fraction tends to retrograde more because of the ease with which the straight, needle-like members can pack together. The branched portion resists ready alignment and remains relatively more stable and fluid.

Reduction in viscosity of starch solution and its tendency to retrograde is necessary to produce a suitable adhesive for paper coating.

Starches are modified by physical and chemical means to render them acceptable for the manufacturing process and product quality. Application of heat and pressure, enzyme action, chemical treatment, and formation of derivatives are all forms of modification that fracture the molecule and lower the molecular weight.

Enzyme Conversion. The most economical and practical form of modification of raw starch in the coating plant is enzyme conversion. However, the cooking requires careful regulation of mixing, pH, time, and temperature to assure reproducibility. Therefore, an appreciable investment in satisfactory equipment is necessary. Enzymes are orgainc substances produced by living matter; they have the ability to act as biocatalysts and to accelerate reactions. Alpha amylase is an enzyme capable of liquefying starch without producing large amounts of sugar, thereby preserving the bonding ability of the starch. Enzyme action, which causes excessive hydrolysis, produces low-molecular-weight maltose, which does not possess any pigment-bonding ability.

Raw starch used for enzyme conversion is buffered by the manufacturer to optimum pH for satisfactory conversion. Slightly premodified starches are now available from the supplier to allow conversion with less enzyme, more uniformly reproducible results, and more stable coatings. Increasing concentration of enzyme on a given starch produces lower viscosity and decreasing bonding ability. Similarly, the water-holding ability of the system decreases as the fractionation progresses.

Starch can also be converted with enzymes in the presence of the pigment slurry to produce lower viscosity and better bonding than can be obtained by separate conversion. Usually the clay must be treated with approximately 1.0% of sodium silicate to avoid excessive absorption of enzyme. The pigment is first dispersed with a polyphosphate in water containing the silicate. The pH is adjusted to 7.5, the starch is added, and finally the enzyme is introduced. Heat is applied until gelatinization occurs and the temperature is held for about 30 minutes. The enzyme is inactivated by adding copper sulfate or by heating to boiling.

Recently, high-pressure jet cookers have been introduced to convert common starch either alone or in the presence of pigment. Continuous conversion is attained by momentary exposure to high-pressure steam at 320°F, followed by rapid reduction in pressure, which literally explodes the starch granule and disintegrates the pigment. Low viscosities and excellent bonding strength are obtained in this process without the use of enzymes.

Chemical Modification. Chemical modification and derivatization of the common starch is carried out in the starch manufacturer's process rather than in the coating plant. Oxidized starch is produced by treating a common starch suspension at 30% solids with a hypochlorite solution in the presence of sodium hydroxide at a pH of 9.0. Approximately 8% available chlorine based on the dry starch is required to obtain the desired amount of oxidation. When the reaction is complete, sodium bisulfite is added to adjust the pH, the material is filtered, washed free of salts, and dried. The degree of oxidation is controlled by varying the pH, time, temperature of reaction, and, amount of hypochlorite.

Oxidized starches are white, gelatinize at low temperatures, require a short cooking cycle, have good bonding strength, and result in low viscosity solutions. Oxidation introduces carboxyl groups into the starch molecule, yielding a product with some of the properties of an organic acid. During the oxidation process, the excess caustic soda replaces most of the carboxyl groups with sodium.

Oxidation is also carried out on the acid side with periodic acid. The chemical action is believed to involve conversion of the glycol groups to dialdehyde starch. The oxidation product can be used as an insolubilizer for protein or other starch adhesives.

Dextrins. Dextrins are produced by roasting common dry starch at high temperatures with or without catalysts. British Gums are obtained by roasting neutral or slightly alkaline starches at high temperature. They have relatively high viscosity, low solubility in cold water, and are tan in color. Canary dextrins are obtained when the starch is roastedin the presence of small amounts of hydrochloric acid. The typical canary yellow color, lower viscosity, and higher solubilities characterize this product. The canary dextrins are too highly converted to be used for coating adhesives. White dextrins are produced by roasting with larger quantities of hydrochloric acid as a catalyst at lower temperatures. A wide range of products is available from low to high viscosity and they are widely used in coating adhesives.

Derivatized Starches. The numerous reactive hydroxyl groups combine with other substances to form chemical derivatives of starch. Ethers, esters, acetals, formals, urethanes, and metallic alcoholates are formed by substitution of the hydrogen atom in the hydroxyl group. Replacement of the entire hydroxyl group by halogens has also been accomplished. Commercially available derivatives for coating adhesives include include a starch acetate and hydroxyethyl starch. The starch acetate is produced by treating starch with acetic anhydride and acetic acid.

Derivatized starches have less gelling tendency, improved water retention, more stable viscosity, improved film-forming ability, and are more reactive with latices than normal starches.

Other Natural Adhesives.

Corn protein is a naturally-occurring high polymer, obtained from the gluten layer of the corn kernel. In milling wet corn, the gluten is separated by isopropyl alcohol extraction from the starch and oil-bearing germ contained in the fibrous hull. Corn protein may be applied to paper from alkaline water solutions as a pretreatment to reduce porosity and to improve hold out of subsequent coatings. It is also used to impart resistance to grease and scuffing.

Alginates are the salts of alginic acid and are hydrophyllic colloids prepared from different types of seaweeds. Alginic acid is a mixture of mannuronic and guluronic acid. Alginates based on high mannuronic acid content are better film formers and have favorable rheology for use in surface treatment of paper. Alginic acid itself is insoluble in water, whereas the salts of sodium, potassium, magnesium, ammonium, and lower amines are soluble in cold and warm water. Insoluble alginate films and pigmented coatings are obtained by the action of polyvalent cations. Such coatings are water wettable, yet water insoluble— an important characteristic in specialty coated paper for lithographic masters.

Synthetic Latices

Colloidal dispersions of synthetic polymers in water are used as paper coating adhesives. A wide choice of individual polymers or blends of different polymers is available to obtain a broad range of physical and chemical properties. The dispersions became known as synthetic latices because of their resemblance to natural rubber latex, which was used in early work to obtain an elastic coating adhesive. The natural rubber proved to be unsuitable because of discoloration and loss of adhesion with age, attributed to oxidation of the rubber. The emulsion polymerization process has yielded stable dispersions of styrene–butadiene, butyl rubber, acrylates, butadiene–acrylonitrile, vinyl acetate–ethylene, vinyl acetate, and butadiene–methyl methacrylate, which are of interest to the paper coater. These latices possess properties that improve performance of the coating mix and impart desirable characteristics to the coated paper. The potential combinations of interest are manifold within the synthetic polymer class alone. In addition, these materials are also blended with natural or synthetic alkali-soluble binders.

Functions of Latices. Considerable investigation into the merits of each synthetic polymer dispersion has revealed that these materials have much in common regarding physical properties, methods of use, and effect on final characteristics. A latex is generally used as a partial substitute for a natural adhesive. The blend causes reduced viscosity and decreased water retention in the coating. Preparation of coatings with higher solids is then possible and the combination of less water and easier water release promotes higher drying rates. Higher output of dry material from the mixing equipment and production at higher speed increases the capacity of the machinery.

Being thermoplastic elastomers, all latices used by the paper coater impart certain desirable properties to the finished paper, the extent of the effect depending on the specific polymer involved. All latices improve resilience, minimize dust, minimize the calendering, increase the smoothness, and provide flexibility, which increases folding endurance. The printer experiences better reproduction of images, reduced sensitivity to moisture, less curl, better holdout of ink and varnish, stronger coating bond, and greater resistance to scuffing.

A latex selection is generally based upon consideration of the requirements of the finished coated paper, the bonding strength, compatibility with available processing equipment and formulations, and economic factors.   Differences in particle size distribution, foaming tendency, odor, and tendency to block are basic qualities that are also examined.

Styrene–Butadiene Latex.   Styrene–butadiene latex possesses many properties that make it suitable for paper coating.   These desirable features and relatively low cost, plus the early introduction as coating adhesives, have given styrene–butadiene wide acceptance.   The standard grade contains 60 parts of styrene and 40 parts of butadiene, a combination that optimizes the benefits of each polymer.   Variations in this proportion lead to major changes in the properties of the coated paper. A greater proportion of styrene yields higher gloss, less flexibility, decreased moisture resistance, and poorer bond.   Although many styrene–butadiene latices are available, a representative description is as follows:

Percentage solids	48 ± 0.5
Specific gravity	1.005
Weight per gallon, lb	8.4
Pounds of solids per gal	4.0
pH	10.5 ± 1.0
Average particle size, μ	0.2
Styrene–butadiene ratio	60/40
Viscosity, cP @ 25°C	28-30

Outstanding compatibility with natural and other synthetic adhesives permits blending of styrene–butadiene to obtain properties superior to those of a single adhesive.   Styrene–butadiene copolymers possess some unsaturation, a property that causes the water resistance and bond to improve by oxidation during normal ageing.

Styrene–butadiene blends well with casein, soybean protein, starch, and polyvinyl alcohol.   Replacement of 25–50% of a natural adhesive with latex gives the desired improvements in coating operation and finished properties.   Experience has shown that a pound of styrene-butadiene latex can be substituted for a pound or more of natural adhesive.   The blend gives a lower viscosity at a higher solids concentration than can be obtained with the natural adhesive alone.

Excellent grease-resistant coatings are obtained with the new low-odor, styrene–butadiene latices.   Blends of methyl cellulose and styrene–butadiene latex also provide excellent films with grease-resistant properties.   The blend reduces the tack of the all-latex film.

Acrylic Latex.   Acrylic polymers produce films that are clear, free from odor, and remarkably stable to heat and light.   Acrylic latex emulsions have unusual mechanical and chemical stabilities and can be formulated to give coatings of high solids and low viscosity. A wide range of properties can be obtained by suitable copolymerization of the acrylic esters. The methyl methacrylate polymer is hard and tack-free, whereas the butyl acrylate is soft and tacky. The properties of the polymer vary with the nature of the ester group as well as the degree of substitution of the acid group. The polymer becomes softer, tackier, and more extensible as the ester goes from methyl to ethyl or butyl.

Acrylic emulsions properly compounded with pigment or hard natural adhesives give coatings with excellent resistance to grease and solvents. This feature has contributed to the extensive use of acrylic emulsions in paperboard applications in the food industry. Coatings containing acrylic emulsions develop instantaneous wet-rub resistance without the need for ageing, as is the case with butadiene–styrene latex.

Nitrile Latex. Copolymers of butadiene and acrylonitrile dispersed in water are known as nitrile latex. The emulsions of interest to the paper coater contain from 20—50 % acrylonitrile. The acrylonitrile polymer contributes adhesion and oil resistance, whereas the butadiene possesses the rubber properties of softness, flexibility, and thermoplasticity. The film properties have found special application where exceptional oil or grease-resistant coatings are required.

Butyl Rubber Latex. Butyl rubber latex contains a polymer that is much more elastomeric and less resinous than other latices. Consequently coatings based on it are much more compressible and flexible. The latex is odorless and films or coatings have excellent stability of color and resistance to ageing. A butyl latex-based coating is as compressible as the paper itself, allowing small surface irregularities to be flattened out under the pressures in letterpress printing.

Polyvinyl Acetate Latex. Polyvinyl acetate latex is prepared by the polymerization of vinyl acetate in water. Other monomers are often copolymerized with vinyl acetate to achieve special properties. The homopolymers of vinyl acetate produce relatively hard films. Improved flexibility has been obtained by copolymerization of 50 % vinyl acetate and 50 % ethylene. Average particle size and distribution of the polyvinyl acetate particles affect the appearance, wet strength, and scrub resistance.

Fine particle size is associated with clear, glossy films, whereas larger particles produce less transparent, duller films. Paper coatings based on polyvinyl acetate have high pick resistance, brightness, and gloss, and do not yellow with age. The latex is compatible with starch, protein, casein, and polyvinyl alcohol.

Polyvinyl acetate dispersions offer a wide range of heat-seal characteristics that vary according to the molecular weight and particle size of the polymer. The low-molecular-weight material heat-seals at about 140°F, whereas the high-molecular-weight product seals above 250°F. The basic heat-seal characteristics are also modified with liquid plasticizers to lower the sealing temperature and with solid modifiers of high melting point to raise it.

Vinylidine Chloride Polymers. Vinylidine chloride polymers dispersed in water and stable and fluid at 60 % solids, are available to the paper coater. Films of polyvinylidene chloride polymer possess unusual barrier properties, particularly for grease and solvent holdout and for gas and vapor resistance. The water-base systems are capable of forming clear continuous films at room temperature without the aid of a plasticizer. The film also produces high gloss, mar resistance, good blocking resistance, has the ability to be heat-sealed face to face and to be glued face to back, and can be readily printed upon without special treatment. The emulsion is applied to paper and paperboard surfaces with no modification or compounding. The coating is applied by coating machinery of the air-knife, metering-bar, wire-wound-rod, size-press, and trailing-blade types.

Water vapor transmission rates below $1g/100in.^2 \cdot 24hr \cdot 0.5mil$ are attainable with clear polyvinylidene chloride films. Oxygen permeability near $0.5cm^3/mm \cdot cm^2 \cdot sec \cdot cm$ Hg at 25°C is also characteristic of a a continuous film. Excellence as a barrier to gas and vapor, resistance to grease, inert nature, stability, and heat-seal characteristics have been successfully applied in the protective packaging field.

Polyvinyl Alcohol

Polyvinyl alcohol is a water-soluble synthetic resin produced by the hydrolysis of polyvinyl acetate. Manufacture of a series of polyvinyl alcohols that differ sharply in physical properties is possible by varying the operating conditions. The degree of polymerization and the percentage hydrolysis are the principal factors governing these characteristics. Variations in the chain length or degree of polymerization of the intermediate polyvinyl acetate influence the molecular weight of the resulting polyvinyl alcohol. Molecular weights of polyvinyl alcohol are difficult to measure with accuracy but they are directly proportional to the viscosity of the solutions. Within a given range of viscosity (range of molecular weights), several grades of polyvinyl alcohol can be produced by varying the degree of hydrolysis of the polyvinyl acetate. Those hydrolyzed 99% are called "completely hydrolyzed," whereas those hydrolyzed to a lesser degree are known as "partially hydrolyzed." As the degree of hydrolysis increases, tensile strength and resistance to water and other solvents also increase, although adhesion properties, dispersing ability, and flexibility decrease. As the molecular weight increases, the viscosity, water resistance, dispersing power, tensile strength, and flexibility also increase. An optimum balance of properties can be obtained for paper coating by the choice of the appropriate grade.

Polyvinyl alcohol is available as a white powdered resin that can be readily dissolved in water and formulated into paper coatings. It is well suited as a binder in pigmented coatings where the ratio of binder to pigment is unusually low and the aqueous dispersions flow smoothly, producing tough white coatings with improved printing qualities. With kaolin clay alone as the pigment, good results can be obtained with 2—5 parts of polyvinyl alcohol to 100 parts of clay. Even though polyvinyl alcohol is water soluble, resulting coatings can be treated to give satisfactory resistance to water. The lower content of binder results in improved brightness, opacity, gloss, and smoothness. Blending the polyvinyl alcohol with butadiene-styrene or acrylic latices further improves the optical properties, reduces the viscosity, and lowers the cost of the coating. The lower adhesive content allows easier calendering, which preserves the sheet compressibility.

Lowering the adhesive level also leads to a marked increase in brightness and opacity with considerably less coating. Four or 5 pounds of coating bound with polyvinyl alcohol are equivalent to 7 pounds of starch coating. This choice represents a 30% reduction in the coating weight and results approximately in a 10% reduction in basis weight of the finished paper.

The polyvinyl alcohol can be made insensitive to water by adding high-purity glyoxal urea formaldehyde or trimethylol melamine to the coating. The combination of polyvinyl alcohol, latex, and curing agent results in a high-quality sheet that can be successfully printed by offset lithography.

Alkali-Soluble Polyvinyl Acetate

Copolymers of vinyl acetate and other monomers that contain carboxyl groups are soluble in dilute alkalis. The presence of the carboxyl group on the polymer chain also improves adhesion to paper surfaces. Low, medium, and high viscosity grades with various degrees of carboxylation are available. The resins dissolve readily in ammonia, sodium carbonate or bicarbonate, and sodium hydroxide to form clear low-viscosity solutions at 15—20% solids. Viscosity of the solution is most stable in the pH range 8.5—9.5 at solids below 20% with ammonia as the alkali.

The adhesive solution can be blended with proteins, starches and latices to obtain satisfactory pigmented coatings for applications in the graphic arts.

Table 4-2 is an attempt to rank the various classes of adhesives qualitatively in terms of prominent properties of interest to the formulator of paper coatings. Small variations may exist within a given classification, particularly when the influence of base paper, coating methods, drying, and other operating factors are considered. The assigned number designates the order of preference in each class.

TABLE 4-2

Qualitative Preference Ranking of Adhesives

	Bonding Ability	Optical Properties	Flow Properties	Cost
Polyvinyl alcohol	1	1	3	5
Latices	2	3	1	4
Proteins	3	2	2	3
Starch derivatives	4	4	4	2
Modified starches	5	5	5	1

White Pigments

Mineral pigment, the predominant consituent of a coating layer, has a controlling effect on the behavior of the wet coating mix and on the properties of the finished product. The ability of a coating to perform well in the process, to provide a smooth and receptive surface, to mask the base stock, and to impart suitable appearance depend on the chemical and physical properties of the pigments.

Physical Properties of White Pigments

Important pigment properties as they relate to the various stages of coating preparation are the ability to disperse and yield low viscosity with appropriate flow characteristics for the intended coating method. The material behavior has an effect on the choice of mixing equipment, material handling, and design of pipes and pumps. Aqueous pigment suspensions at 72% by weight and pigment-adhesive systems at 45-65% must possess appropriate flow properties that enable efficient agitation, minium settling, easy transport through pipes, and ease of screening.

Fluid coatings at high solids content depend on good dispersion in which the vehicle acts as a lubricant between particles and allows them to move freely over each other. Pigments vary in their tendency to aggregate and in their resistance to dispersion. Different chemical composition, particle size and shapes, and the amount of soluble materials present may cause variations in the forces that hold particles together.

These forces are generally weaker in kaolin clays than in titanium dioxide or calcium carbonate and must be overcome to disperse and suspend the pigment satisfactorily. The time and intensity of mixing action required to disperse the different pigments varies with the magnitude of these aggregative forces. A combination of mechanical force and the chemical action of dispersing agent is usually required to break down the pigments. As the intensity of mixing increases, the degree of dispersion improves. Greater input of energy and more efficient utilization of the mixing energy can be obtained by increasing the speed of the mixer and by increasing the concentration of the pigment.

Brightness. The whiteness or brightness of pigments is determined by chemical composition and purity. For kaolin clay, the mineral is chemically bleached to remove color that is caused by iron compounds. The brightness of the final coated paper is determined by the brightness of the base stock and the brightness and hiding power of the pigments in the coating. Base stock with low brightness requires more coating to mask it. The brightness of the coated sheet is the same as the ultimate brightness of the coating when the base stock is completely masked. This point is seldom reached unless the brightness of the coating and the base stock are similar.

Refractive Index. The difference in refractive indices of the various coating ingredients determines the hiding power or opacity of the coating. The refractive indices of adhesive and air in the coating are low relative to those of the pigments. A pigment with a high refractive index increases the difference between it and the other ingredients and thereby results in high opacity. The most opaque coating would consist of pigment and air because it would produce the greatest difference in refractive indices. As air is replaced by an adhesive with a refractive index closer to that of the pigment, the opacity of the coating is lowered. Therefore, as the adhesive content increases and air is displaced, a less opaque coating is obtained. This action is similar to calendering, whereby the opacity decreases as the increasing pressure expels air from the coating voids.

Configuration. The size, shape, and arrangement of pigments in a coating influence the physical and optical properties of the surface. Combinations of shapes and sizes without limit are possible. Particles can be assembled in loose clusters or can be tightly packed together. Kaolin clays function as satisfactory coating pigments mainly because they consist of hexagonal plates with diameters below $2\,\mu$. These thin plates, 1/10 of their diameter in thickness, assume positions parallel to the substrate. This regular arrangement exposes a uniformly flat surface that is capable of accepting a gloss and that can be leveled to provide a smooth structure. Small clay particles are usually brighter than coarse ones down to a diameter of $0.25\,\mu$. Clay decreases in brightness and opacity below this size.

Calcium Carbonate

Calcium carbonate crystals exist as cubes or needles. The multiple faces of the crystals form numerous surfaces for reflection or refraction of light. Consequently, most of the light is reflected outward rather than passing through the coating layer. The combination of many light-active surfaces and extremely small particle size creates high opacity in thin layers. The blend of plates, cubes, and needles can be varied to provide a variety of structures that differ in their ability to absorb ink or varnish. Extremes of holdout and absorption are attainable by selection of proper kind and quantity of different physical forms. Particle shape also influences the flow characteristics of wet coating; perfect spheres would be most desirable. Without good flow an uneven surface on the coated sheet of paper can result.

The quantity of adhesive required to bond pigment particles to one another and to the base paper varies for each pigment. The chemical composition, surface area, and volume of voids between particles determine the adhesive demand. In the case of calcium carbonate, the needle structure assembles with large and numerous voids that trap adhesive. As the mixing action proceeds and the cylindrical particles are fractured and closely packed, the requirement of adhesive decreases.

Kaolin Clay

The paper coating clay used for general printing consumption is known chemically as a complex hydrous aluminum silicate and is hard to characterize. Clays vary in chemical composition and colloidal behavior. A better understanding of these materials has been made possible by differential thermal analysis, the Coulter counter, the centrifuge, the electron microscope, and X-ray diffraction apparatus. The basic building units of kaolin clay are probably few and simple. The differences observed in coating clays have been attributed to the spatial arrangements of the crystal lattice. A pure kaolinite is composed of one molecule of alumina, two of silica, and two of water. Halloysite results if the water content is increased to four molecules. Variations in the decay process of minerals can produce different ratios of alumina to silica to form different minerals.

Origin of Clay. Kaolinites are formed by the weathering of mica or feldspar. Feldspar also weathers to mica, which further decomposes to kaolin clay. Well-formed kaolin crystals result from the direct weathering of mica. Less perfect crystals occur when kaolin is derived from mica that resulted from the decomposition of feldspar. This mechanism of natural formation determines the crystal structure and is typical of domestic clays found in Georgia. The secondary deposits are free from abrasive grit and contain a large proportion of thin hexagonal plates.

Clays found in England are termed primary clays because they are located near the parent rock from which they are formed. The English clays are generally brighter and bluer than secondary-deposit domestic clays.

Paper coating clays are delivered dry or as a wet slurry. The dry clay is supplied in the form of finely divided powder, acid lump, or spray-dried, hollow, spherical droplets.

Recent developments in domestic clay have made available fine-particle grades that are all finer than 2μ. High brightness, high gloss, and superior varnish holdout are characteristic of these materials. The lower particle size permits a high gloss with less calendering which in turn minimizes loss in bulk and resiliency.

Clays beneficiated by special flotation processes yield brightness values of 90 and above. The flotation technique removes colored impurities from standard domestic minerals.

A delamination process has made possible a kaolin structure of thin platelets with large diameters. The absence of contaminants on the newly exposed delaminated surfaces provides clay with brightness at 90 to 91. The physical structure causes leafing of the platelets, which results in good solvent and ink holdout.

Classification of Clays. A general classification of clay grades is made according to particle size and brightness. Table 4-3 describes the major differences in the three grades commercially available.

TABLE 4-3

Properties of Coating Clays

Particle size	No.1	No.2	No.3
(% less than 2μ)	90-94	80-83	65-73
Brightness G.E.			
English	91-92	90-91	89
Domestic	86-88	85-88	84-87

Precipitated calcium carbonates are second only to clay in total volume of pigment consumed in paper coating. Although ground limestone can be used in some operations, when uniformity of particle size and shape are considerations, precipitated calcium carbonate is most frequently employed. Differences in quality exist even among man-made products, causing some to be more suitable than others, depending on the application.

Preparation of Calcium Carbonate. Natural limestone in lump form is roasted in kilns to drive off carbon dioxide and undesirable organic impurities. The remaining calcium oxide is slaked with water to form calcium hydroxide (commonly known as milk of lime), which is screened to remove coarse particles and impurities. The three common methods of making precipitated calcium carbonate diverge from this beginning.

One procedure, known as the carbonation process, is to bubble carbon dioxide from a coke burner through the milk of lime suspension:

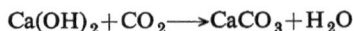

$$Ca(OH)_2 + CO_2 \longrightarrow CaCO_3 + H_2O$$

In the "lime soda process," soda ash is reacted with milk of lime, producing calcium carbonate and caustic soda:

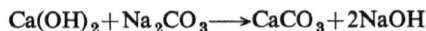

$$Ca(OH)_2 + Na_2CO_3 \longrightarrow CaCO_3 + 2NaOH$$

Difficulty in removing all traces of alkali from the precipitated compound places limitations on this method.

A two-stage, salt-formation method produces a grade of greater purity. The milk of lime is treated first with ammonium chloride, forming a solution of calcium chloride and ammonia in water:

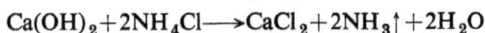

$$Ca(OH)_2 + 2NH_4Cl \longrightarrow CaCl_2 + 2NH_3\uparrow + 2H_2O$$

The ammonia is driven off by heat and the insoluble inert solids are readily removed by filtration. The crystal-clear solution of calcium chloride is treated with a filtered solution of sodium carbonate to yield ordinary table salt in solution and a precipitate of calcium carbonate:

$$CaCl_2 + Na_2CO_3 \longrightarrow CaCO_3 + 2NaCl$$

The sodium chloride is washed away, leaving a pure form of calcium carbonate. Particles produced by the latter method are cubic in shape, whereas the former reactions yield acicular particles.

Properties of Calcium Carbonate. The areas and methods of practical application are determined by the properties of the compounds rather than the generic category of classification of origin. The properties, not the result of chance, are determined by rigid controls throughout the manufacturing steps of all procedures described. The proportions of reactants, rate of addition, degree of agitation, and reaction temperature influence particle size. Control of shape and size affects the total surface area, which largely determines absorption of water and oil and density of packing. Attaining a high degree of purity is essential to reduce abrasiveness.

Calcium carbonate that is pure, white, grit-free, nonabrasive, properly shaped and packed, inert, and oil absorbent contributes useful values to coated paper through properties that are essentially physical rather than chemical.

Calcium carbonates are chosen for use largely because of their high brightness, oil absorption, and tendency to produce a matte finish. Their needle and rhombic configuration also tends to produce a porous coating structure in combination with kaolin hexagonal plates. This structural design promotes the escape of moisture from the coated paper in the heat-set ovens on web offset printing without blister or rupture. The particular orientation of pigment also contributes toward the lower gloss and higher ink absorption.

Pertinent properties of calcium carbonate appear in Table 4-4.

TABLE 4-4

Calcium Carbonate Pigments

	Ground calcium carbonate	Precipitated form, carbonation process	Precipitated form, double-salt process
Brightness G.E.	96	98	100
Crystal form	Calcite	Aragonite	Calcite
Shape	Multiple	Needle	Cube
Oil absorption, lb/100 lb pigment	14-17	40-60	30-32

Titanium Dioxide

Titanium dioxide pigment is valuable because of its properties, particularly the combination of high refractive index with optimum particle size. Many processing methods for the manufacture of this pigment result in materials with slightly different properties.

Preparation of Titanium Dioxide. The most common process for manufacture of titamium dioxide involves the reaction of sulfuric acid with ilmenite ore, which yields iron and titamium sulfates. The chloride process involves the chlorination of titanium ore in the presence of carbon to produce titanium tetrachloride. The ore is converted to ferric chloride with the carbon acting as a reducing agent.

Properties of Titanium Dioxide. Titanium dioxide is particularly suitable as a pigment because it is stable. Higher and lower oxides are unstable and the latter compunds are dark in color. Extreme insolubility in acid or alkali over a wide temperature range indicates the chemical stability of the compound. It is unaffected by the hydrogen sulfide and oxides of sulfur usually formed in the atmosphere. This extreme stability is believed to be the result of the strong bond between titanium and oxygen, which possibly derives from the protection each titanium atom is given by the six oxygen atoms in the crystal structure.

The three known crystal types are anatase, brookite, and rutile; anatase and rutile are tetragonal, brookite orthorhombic. The difference in spatial arrangement of the lattice and the degree of atomic packing are considered the bases for differences in physical properties. The more compact structure of rutile accounts for its higher refractive index. Anatase at 2.52 and rutile at 2.76 possess the highest known refractive indices for white pigments.* This property determines the light-scattering ability of finely divided titanium dioxide with consequent high reflectance and opacity. Anatase displays a more bluish white hue; rutile tends to have a yellowish cast.

Titanium dioxide (at 25—27 cents per pound) is usually blended with other pigments to optimize its benefits at minimum cost. The greatest rate of improvement is obtained when the titanium dioxide constitutes 5—25% of the total pigment.

Rutile titanium dioxide displays high absorption for ultraviolet light. As a consequence, it tends to reduce the effectiveness of fluorescent dyes, and such combinations should be avoided.

The rutile form with higher refractive index is more efficient for applications that require waxing, oil impregnation, or varnishing. Whenever these requirements are not important, the anatase type is

---

* Refractive indices for other white pigments are: clay, 1.55; calcium carbonate, 1.56; talc, 1.57; calcium sulfite, 1.57; barium sulfate, 1.64; titanium–barium composite, 1.91; titanium–calcium composite, 1.98; blanc fixe, 1.64; and zinc sulfide, 2.37.

more economical to use. The specifications for the finished product and the bodystock involved strongly influence the choice of form to be applied. (see Table 4-5).

TABLE 4-5

Characteristics of Titanium Dioxide Coating Pigments

	Anatase	*Rutile*
Bright G.E.	96-98	97-98
Refractive index	2.52	2.76
Specific gravity	3.9	4.2

Barium Sulfate

Barium sulfate is attractive as a paper coating pigment because of its superior whiteness and extreme insolubility. The common trade name for refined barium sulfate is blanc fixe; sometimes called baryta. Its high purity, density, inert character, and ability to accept tinting dyes make it especially suitable for a subcoating beneath photographic emulsions. These same desirable properties are useful in special coated papers for converting letterpress proofs to lithographic printing plates.

Manufacture of Barium Sulfate. Barium sulfate occurs in the United States in the form of crude barite, which is mined, ground into small particles, and bleached with acids. Small amounts of the bleached ground barite are used as matting agents in special dull-coated papers.

The impure barium sulfate (barite) is purified and recrystallized to produce blanc fixe. A soluble barium salt is precipitated with a soluble sulfate to form pure blanc fixe. The pure barium sulfate is sometimes mixed with zinc sulfate to produce a composite pigment known as lithopone.

Properties of Barium Sulfate. The optimum particle size for barium sulfate is between 0.4 and 0.6 μ. Particles smaller than 0.4 μ produce a higher gloss but with a loss in whiteness. Particles larger than 0.6 μ produce a lower gloss with higher whiteness and higher opacity. At the appropriate particle size, barium sulfate requires less adhesive for satisfactory bonding than that required for coating clays.

Satin White

Satin white is formed by the interaction of slaked lime and aluminum sulfate.

Composition. Most manufactures and users concur that the resulting white pigment is calcium sulfoaluminate, $3CaO \cdot Al_2O_3 \cdot 3Ca SO_4 \cdot 31H_2O$. The pigment is crystalline in nature in the form of minute needles. These crystalline needles are believed to be held together by water bridges. Drying the pigment for extended periods transforms the structure to an amorphous powder.

Functions. Satin white has a bright white color, produces a high gloss on calendering, and contributes toward waterproofing the coating when used with casein. It is generally mixed with other pigments, accounting for 40-50% of the total pigment composition. Occasionally it is used in amounts (3—5%) to impart water resistance to the coating. In small quantities it has little effect on the gloss or color of the coating. Satin white requires much larger quanities of adhesive than other coating pigments.

Properties. Satin white pastes are viscous, sticky, and difficult to handle. They can be made fluid by the addition of small quanities of gum arabic or sugar. As little as 10% casein based on the air-dry pigment, in combination with high input of mechanical mixing, is also useful to fluidize satin white.

Satin white loses about 28% of its air-dry weight on heating to 100°C and 31% on heating to 140°C. In normal drying of coated paper, this water of crystallization is retained in a coating layer and is responsible for its ease of calendering and high gloss. Upon total ignition, satin white loses 48% of its air-dry weight.

Recent refinements in the control of manufacture have caused a revival of interest in satin white as a pigment in coated paper because of demands for whiter and higher finish papers.

Talc

In recent years, newly developed talc with exceptional softness, purity, ultrafine particle size, and nonabrasiveness has been made available to the paper coater.

Composition. The new grades have particle sizes all below 5 μ and a Mohr hardness of 1.0. Mineral talc is hydrous magnesium silicate with composition as follows: $3MgO \cdot 4SiO_2 \cdot H_2O$. Talc in its purest form contains 31.9% MgO, 63.4% SiO_2, and 4.7% H_2O. Impurities such as calcite, quartz, and dolomite increase the abrasiveness and cause dark color.

Unlike kaolin clay, crude talc requires a large input of mechanical energy to reduce the particles to the same fineness and surface area as those of coating clays. Equipment needed to accomplish this grinding has only recently been available. Therefore, the newer talc pigments consist of much finer particles than those previously offered to the coating industry.

Talc varies in shape depending on the source and degree of purity. The basic geometrical shape of the particle is retained as the particle size is reduced. The purest grades occur in platelike or micaceous structure. Less pure forms occur as fibrous or circular shapes. The pure variety, reduced to fine particle size platelets, has excellent calendering properties and good ink holdout.

Use of Talc. Ultrafine talc powder contains up to 97% by volume of entrapped air which, in combination with the hydrophobic nature of talc, makes wetting with water difficult. Nonionic wetting agents of the polyoxyethylene–polyoxypropylene type are effective in wetting talc. A combination of agents is required to wet and disperse talc effectively. Nonionic wetting agents combined with polyphase dispersing aids usually accomplish the necessary results. Talc and water slurries at 60% solids are possible with 1.3% $(K_3PO_4 \cdot 2KPO_3)$ in combination with 2.5% nonionic wetting agent.

Zinc Oxide

Major advances have been made in applications of specialty coated paper that involve properties of zinc oxide such as semiconductivity, luminescence, photoconductivity, and photochemical effects.

Preparation and Particle Size. Zinc oxide, composed of one atom of zinc and one of oxygen, has many chemical and physical properties that cause it to display semiconductor and photoconductive abilities. It is produced by the American and the French processes. The oxide is prepared directly from the zinc sulfide ore in a continuous operation by the American process. The metalic zinc is produced first; then it is converted by vapor oxidation into zinc oxide by the French process.

Zinc oxide is available in fine particle sizes; the average numerical diameters range from 0.15 μ in finer grades to 1.0 μ in the coarser grades. Fine particle size and high specific surface are complimentary, indicating a high order of chemical reactivity. This property is utilized in the formation of zinc caseinate in casein-bound coatings to improve wet-rub resistance.

Properties. Zinc oxide is relatively opaque to ultraviolet light, a property that impedes the passage of these light rays of short wave length when the pigment is used with elastomers in coatings. Zinc oxide protects these organic materials from attack and deterioration by actinic light. The following table gives a comparison of ultraviolet transmission of zinc oxide and other pigments.

Pigment	% transmission of ultraviolet light at 3600 Å
Zinc oxide	0
Titanium dioxide	18
Calcium carbonate	68

Surfaces with zinc oxide pigment possess electrostatic, semiconductor properties. As early as 1911 tests showed that oil paints containing zinc oxide are more rapidly destroyed by light than those containing other inorganic substances. A layer of photoconductive zinc oxide pigment properly compounded has high electrical resistance in total darkness but a relatively low resistance when illuminated. The layer of zinc oxide coating can be made light sensitive if it is uniformly electrically charged with ions usually formed by corona discharge operating at high voltage but low current. This unique property makes zinc oxide suitable for copying by electrophotoconductive means. Zinc oxide prepared by the French process produces the purer grade suitable for use in copy paper.

Diatomaceous Silica

Diatomaceous silica is characterized by its chemical inertness, minute porosity, great bulk, and high absorption capacity. It is also known as infusorial earth and is formed from residues of aquatic plants known as diatoms. The particles are extremely thin transparent shells of diverse shapes, and particle size averaging about 10 μ. The natural material can be flux-calcined to produce a white substance at high brightness.

The pigment is usually added to a coating in small proportions to reduce or eliminate gloss. It also provides a noticeable resistance to calender blackening.

Luminescent Pigments

Luminescent pigments are materials capable of luminous radiation of light more intense than that coincident with the temperature of the emitting body. The pigments can be either phosphorescent or fluorescent. Phosphorescent pigments have a useful period of afterglow following exposure to black or visible light. The period of afterglow varies but it may amount to several hours. Fluorescent pigments emit visible fluorescent light during exposure to a source of light radiation, but do not show an afterglow when the source is removed. Fluorescent pigments are subdivided into black-light- and daylight-fluorescent compounds. Phosphorescent pigments can be fluorescent under the action of black light.

Coated papers with a novel luminescence are of significant interest for applications as display posters, decorative paper, gift wrapping, and packaging labels.

Zinc Oxide Type

Phosphorescent pigments and black-light fluorescent pigments consist of pure activated zinc oxide in combination with sulfides of zinc and cadmium. Sulfides of calcium and strontium are also sometimes used. The latter combination has a longer afterglow, usually six to twelve hours.

Daylight-Fluorescent Pigment

Daylight-fluorescent pigments are either organic dye and resin combinations or straight crystalline compounds. The resins are generally vinyl or amine-aldehyde types. Xanthene and acridine type dyes, which are soluble in the resins, are generally used. Usually the concentration of dye is critical, in that excess use of material destroys the daylight fluorescence.

The daylight fluorescent colors have a well-defined reflective hue and also absorb invisible ultraviolet light and convert it into visible light. Brilliant shades of orange, green, buff, and white are developed. The shades can also be modified through the use of other tinting colors.

Black-Light Fluorescent Pigments

The black-light fluorescent pigments develop billiant orange, red, yellow, green, and blue colors when excited by ultraviolet light.

Colored Pigments

A large part of the production of a speciality mill is concerned with colored book papers, labels, bristols, covers, and identification tags. The choice of colorants depends upon the requirements of the finished product. Properties such as bleed, tinctorial strength, brilliance, reactivity, fade resistance, and cost must be considered by the formulator in choosing a particular color combination. In some cases, the entire pigment component may be colored; in others, only a small percentage is required.

Pigments or water-insoluble materials used for coloring water-base mineral coatings are classified as inorganic and organic compounds. Natural and synthetic products occur in each class. Water-soluble dyestuffs, natural and synthetic, are also used where soluble compounds are applicable.

Inorganic Pigments

Natural inorganic pigments are primarily iron oxides and hydroxides. Yellow pigments are derived from ochre, raw sienna, and natural oxide. Browns are obtained from raw or burnt umber. Burnt sienna, pyrites, cinder, and Spanish or Persian Gulf oxides are a source of reds. The natural substances are generally of low strength, irregular size and shape, and contain undesirable hard grit.

Synthetic inorganic pigments include lead chromates ranging in shade from primrose yellow through orange. Similar shades are also obtained with cadmium sulfide compounds alone or in combination with coprecipitated barium sulfate. Browns with yellow, red, or purple undertones can be synthesized as iron oxide compounds. Reds and maroons of various shades are based upon cadmium sulfoselenide and cadmium–mercury compounds. Chrome oxides and hydrated chrome oxides provide a wide range of greens. Phthalocyanine green and blue compounds are used alone and in combination with chrome greens to develop a wide range of hues.

The inorganic pigments, natural and synthetic, are considered fast to light, bleed-proof, and stable to moderate alkali environment. They are used as dry powder or in paste form and respond to dispersion and mixing by means of treatment similar to that for white pigments.

Organic Pigments

Natural organic pigments derived from animals and plants are of little importance in coloring paper coatings. Cochineal and carmine from insects and sepia from cuttlefish are no longer used in paper coating. Similarly, the metal salts of plant extracts have been replaced by synthetic organic pigments.

Soluble organic acid, basic, or direct dyes in combination with selected inorganic white pigments constitute a class of colorants known as color lakes or pulp colors. The insoluble lakes are produced by precipitating various coal tar dyes upon an insoluble base by means of a metallic salt, tannin, or other reagent.

Acid dyes contain one or more replaceable hydrogen ions that form a lake by metathesis with a metallic compound. The hydrogen of a sulfonic acid carboxyl group or phenolic group is usually replaced with an alkaline earth metal. The acid dye might also be absorbed on a hydrous metallic oxide such as a coating clay.

Basic dyes always contain one or more amino groups that can be precipitated by tannic, gallic, phosphotungstic, phosphomolybdic, or phosphomolybdic-tungstic acids. Basic dyes are precipitated with agents that form insoluble complexes with nitrogen compounds. Neutral dyes can be precipitated on inert substances such as aluminum hydrate, barium sulfate, china clay, or calcium carbonate.

Color lakes are available in a wide range of shades, and unlike the soluble dyes from which they are derived they become insoluble in water and considerably faster to light. The basic dyestuffs by themselves are the brightest and tinctorially strongest but are quite fugitive on exposure to light. Acid dyes are somewhat weaker and duller than basic dyes, but are more fade resistant as a class.

Metallic Pigments

Aluminum, copper, and bronze metallic powders are used as coating pigments to obtain special color or functional effects in the coated paper.

Aluminum

Small thin pieces of aluminum in the form of sheet, splatter, or foil are reduced to fine particles in a stamping, ball, or rod mill. The processing in the presence of a suitable lubricant continues until the powder has reached the desired fineness. Most powders used for paper coating pass through a 325-mesh screen.

The flakes of aluminum are relatively flat and thin. The thin flakes have the ability to "leaf" when suspended in water. When applied to paper, some of these flakes are carried to the surface of the coating layer by convection within the wet layer. They are held there by the force of surface tension, forming an almost continuous metal leaf at the surface. This effect produces a metallic luster and a film that approaches that of a laminate of foil paper.

Copper and Copper Bronze

Copper powders are manufactured by electrolytic deposition from solutions of copper salts. Acid copper sulfate is most commonly used. The metal is deposited in the dendritic or crystal form by high-speed deposition.

TABLE 4-6

White Pigments

Pigment	Refractive index	Brightness, % MgO	Range of particle size, μ	Adhesive demand per 100 lb pigment
No. 1 Clay	1.60	88-90	95-98% <2	14
No. 2 Clay	1.55	87-88	90-93% <2	14
No. 3 Clay	1.55	84-86	80-82% <2	13
Calcium carbonate	1.5—1.7	94-97	0.1-0.5	18
Titanium–anatase	2.52	96-98	0.2-0.5	12
Titanium rutile	2.76	97-98	0.2-0.5	12
Blanc fixe	1.65	98	0.5-2.0	18
Satin white	—	—	—	35—50
Silica (diatomaceous)	1.40—1.50	90-92	90% <4	14
Talc	1.57	90-97	80% <5	12
Zinc oxide	2.01	97-98	0.3-0.5	12

Copper bronze powder, formed in individual flake particles, gives better hiding power and greater metallic luster than the dendritic powder. Thin sheets or pieces of metal are stamped or ground in the presence of a fatty lubricant, usually stearic acid. The bronze flakes tend to leaf and produce a surface with a metallic luster in much the same manner as aluminum.

Table 4-6 lists the useful properties of common pigments.

Additives

Coating additives are defined as substances that are added to the wet coating composition to enhance, optimize, or otherwise improve the properties of the coating layer or to avoid operating problems during the mixing coating, drying, calendering, or finishing processes. Chemical compounds that prevent or control foam; provide lubrication; and act as dispersing agents, preservatives, flow modifiers, and insolubilizers are ranked among important additives.

Foam Control

Foam is described as a dispersion of air or other gases in a liquid system. It occurs in bubble sizes in the range from 1—100 μ in diameter and can be distributed in sparse or profuse quantities. The size and number of air bubbles are influenced by the nature of the materials, methods of mixing, and design of coating handling and applicator systems. Control of this phenomenon is extremely important to efficient mill operation and optimum quality of product.

Influential Coating Properties. Coating properties that influence foam activity include interfacial tension, pH, solubility, temperature, viscosity, and degree of dispersion. Usually, low surface tension permits foam to form and high viscosity stabilizes it. Adhesives contribute more than any other coating component toward formation of foam. This phenomenon is especially evident in synthetic polymer emulsions where a foamer is present in the form of wetting or emulsifying agents. High-speed mixing equipment presents conditions more prone to create foam and entrapped air.

Chemical Defoamer. Remove of causes of foam formation related to the operation and the proper selection of materials is preferable to supplying a form of cure. Chemical agents for control of foam include pine oil, silicone emulsions, fuel oils, tributyl phosphate, and higher alcohols such as octyl, nonyl, and tridecyl. Extreme care must be exercised in the use of such materials to prevent disagreeable odor, incompatibilities, and unnecessary costs. The most effective and lasting defoamers are generally the most incompatible with aqueous systems and are prone to produce "fish-eye" defects.

Some foam control agents are used as preventatives or antifoam additives to prevent, inhibit, or reduce the tendency of a system to foam. Silicone types, metallic soaps, polyglycol esters, and phosphate esters have the necessary "lasting power" to prevent buildup of foam even with intensive mixing. They are applied during the preparation of the coating in quantities that range from 2—6% of the adhesive solids.

Other materials such as ether types, alcohols, vegetable oils, and emulsifiable pine oil are capable of combating or "knocking down" foam formed in the system. The duration of this effect is short with such compounds and the system quickly builds back the foam destroyed.

Chemical defoamers can sometimes create quality problems more serious than those caused by the foam itself. Fisheyes, oil spots, wicking, and uneven ink absorbency are among some of the likely defects. Nonchemical defoaming with vacuum systems, sonic energy, and mechanical methods have been introduced in recent years. High frequency sound waves cause vibration to particles in the system and drive the air bubbles to the surface where they are broken. Formation of foams can also be reduced by avoiding cascading and flow surges, by eliminating pipes and pumps that aspirate air, and by maintaining full tanks and machine supply boxes.

Surface–Active Agents

Surface activity has been defined as the strong tendency of a solute to concentrate at an interface. Molecules with certain configurations exhibit surface activity. The molecule has two distinct portions; one has a strong affinity for the solvent, whereas the other is rejected by the solvent. When the repulsion forces are strong, the solute molecules tend to concentrate at an interface, with part of the rejected group not in contact with the solvent molecules. Ordinary soaps behave in this way when dissolved in water; the polar group is attracted by the water and the hydrocarbon chain is rejected. This phenomenon causes a concentration of soap at the water-air and water-oil interfaces.

Numerous compounds have been synthesized and offered to the coating industry for use in wetting, foaming, antifoaming, emulsifications, detergency, flotation, and other applications in water systems.

The process of coating paper is a typical industrial application of wetting phenomena. The transfer of ink in printing and the process of pigment flotation depend upon preferential wetting. When coating is placed on the surface of paper it may spread to cover the surface or it may remain as a series of drops.

Aqueous systems containing soluble surface-active materials wet paper surfaces more easily when the concentration of wetting agents reduces the surface tension below a certain critical value. Surfactants also influence coating mixing and dispersion. Their profound wetting action may enhance the dispersion of pigment or the spreading quality of the coating on the base paper. Small quanities used with discretion at optimum conditions may produce a remarkable improvement in quality.

Surface-active materials can be classified as follows:

1. Anionic compounds in solution, with the long chain negatively charged;
2. Cationic compounds in solution, with the long chain positively charged;
3. Nonionic compounds that do not ionize in solution;
4. Amphoteric compounds in solution, with the long chain either positive or negative, depending upon the pH of the solution.

Anionic Compounds. Common sodium and potassium soaps have been used as surfactants for many years.

Ammonia and amine soaps include morpholine mono-, di-, and tri-ethanolamine and water-soluble alkyl amines. These materials are incompatible with acids and calcium or magnesium salts present in hard water.

Treatment of a petroleum product with sulfuric acid produces alkyl aryl sulfonates—an example of a group of anionic surfactants other than soap. Numerous sulfonates can be dervied from higher molecular weight alkylates, which have more powerful surface-active ability.

Sulfonated vegetable oils such as castor oil, coconut oil, or oleic acid have been widely used for their excellent wetting abilities. They are prepared by the reaction of a fatty alcohol with sulfuric or cholo-sulfuric acid.

Cationic Compounds. Long-chain primary, secondary, and tertiary amines and quaternary ammonium compounds are two general classes of cationic surfactant.

Nonionic Compounds. Heating a fatty acid with an excess of diethanolamine forms a complex fatty alkanolamide mixture that is nonionic.

Amphoteric Compounds. Long-chain amino acids, such as proteins, contain both an anionic and a cationic group, remaining cationic in acid solution and anionic in alkali. Solubility, foam, wetting, and depression of surface tension are at a minimum at pH values near the isoelectric point. In neutral or slightly alkaline solutions they exhibit high foaming action typical of anionic surfactants.

Dispersants

The function of a dispersant is to overcome the attraction that pigment particles in fluid suspension have for each other. Forces that tend to attract particles restrict the coating flow and cause undesirable high viscosity. Dispersing agents ionize when dissolved and form large anions, which are adsorbed on the surface of the pigments. These produce an electrostatic charge on the pigment particle equal to the charge of the adsorbed ion. The magnitude of the charge is dependent on the number of ions adsorbed and the charge on each ion. A concentration of adsorbed ions on the pigment surface causes a redistribution of electrical charges in the system and creates a repulsion between particles of similar charge. Common dispersing agents that perform this task are sodium salts of polyphosphoric acid, such as tetrasodium pyrophosphate, sodium hexametaphosphate, sodium tripolyphosphate, and sodium tetraphosphate.

Clay pigments from different sources vary in their dispersant demand. The quantity required for minimum viscosity ranges from 0.1 to 3.0% of the weight of the clay. Beyond the optimum quantity, addition of more dispersant causes the viscosity to increase. Minimum viscosity is obtained under alkaline conditions. Clay dispersion is less difficult where protein, rather than starch, is present as the adhesive. Starch coatings require larger quantities of dispersing agent. Other pigments such as calcium carbonate, satin white, or titanium dioxide increase the requirement of dispersant.

Hydrophyllic Colloids. Hydrophyllic colloids act as dispersing agents by forming a protective layer of negatively charged colloidal particles on the surface of the pigment. Alkaline solutions of casein and soybean protein provide strong protective action. The properties of suspensions stabilized with proteins parallel closely those of suspensions dispersed with ionic dispersants. Coatings prepared with protein as all or part of the adhesive can be prepared at low solids without additional phosphate dispersant. Approximately 5% protein, based on the pigment solid, is adequate for good dispersion.

Nonionic Colloids. Nonionic colloids absorbed on the surface of the pigment also aid dispersion. The highly-hydrated, adsorbed colloid may be a barrier to flocculation without the electrical repulsion. The action of nonionic starch colloid in dispersing satin white pigment is an exanple.

Polyphosphates. The effectiveness of polyphosphates to disperse pigments varies with pH and the hardness of the water. Polyphosphates with high content of phosphorus pentoxide are more effective because of their greater ability to sequester multivalent ions that interfere with dispersion. Less polyphosphate is required with increasing pH of the clay suspension because hydroxyl ions are also effective in promoting dispersion.

Lubricants

The field of paper coating lubrication can be divided into two categories. The first involves hydrodynamic lubrication and occurs when moving pigment surfaces are completely separated by a water layer. The second is boundary lubrication. Here, there is contact of paper and process machinery where absorbed films of lubricants reduce friction between the surfaces.

Surface Lubrication. Additives that lubricate in the fluid coating phase may improve flow properties and alter the total rheology. Surface lubrication can influence smoothness, finish, printing qualities, pliability, and tendency of the coated paper to dust. Water-soluble soaps, sulfonated oils, wax emulsions and insoluble soaps are used as coating lubricants. Most chemicals in this category possess surface-active properties that effect the surface and interfacial forces of a coating. Consequently, they affect the pigment dispersion, alter the flow properties, and influence the penetration of the coating into the substrate. Surface tensions can affect the way a film of coating splits between the applicator roll and the paper, as well as the leveling properties after the application. Good flow and good leveling are associated with low surface tension.

Soaps. Soluble and insoluble soaps are able to lubricate wet coatings. Sodium and ammonium stearate promote uniform wetting of the application roll and uniform distribution of coating on the paper. Soaps also lubricate the dry coating to assist in improving the gloss and smoothness and to reduce the tendency to dust during calendering. Approximately 1—2 % of lubricant based on the coating solids is used to accomplish the necessary effect.

Wax Emulsions. Wax emulsions are often used as coating lubricants, particularly in friction glazing or high-finish calenders and brush machines. Soft and inexpensive beeswax, emulsified in water, produces excellent results in a coating used for glazing. Waxes are chosen as lubricants if they do not develop oily or transparent areas under heat and pressure. Japan wax or stearic acid made into a soapy emulsion by boiling with borax, ammonia, and water provides excellent lubrication in colored coatings for friction glazing.

Preservatives

Water-base coatings containing protein or fatty acid stabilizers can support microbiological activity and growth. Spoilage leads to production problems, economic losses, and dissatisfaction on the part of the customer. Action of microorganisms can cause poor coating adhesion, poor coating printability, and disagreeable odors.

The type of coating to be preserved dictates the choice of fungicide or bactericide. The same preservative may not be effective against casein, animal glue, starch, or soya protein. Besides the requirements of the coating, the end use of the final product has an influence on the selection. Materials suitable for a soap wrap may not be acceptable for food wrap or contact.

Spoilage Microorganisms. Spoilage microorganisms include bacteria and fungi. These living organisms require basic elements to support life and reproduction; carbon and nitrogen available from organic or inorganic substances are the two most important. Adhesives such as glue, starch, soya, protein, casein, styrene–butadiene, and acrylics are troublesome components. Carbon and nitrogen are available in pigments, additives, and contaminants. The temperature, pH, aeration, and length of storage influence the kind and quantity of microorganisms to be combatted.

Control of Spoilage. Spoilage can be controlled by cleanliness, good housekeeping, and use of chemicals. Regulation of temperature and change in pH may also be helpful in controlling development of bacteria, which normally grow in a pH range 4.0—9.0; the highest activity occurs between 6.0 and 8.0.

Chemical classes of preservatives include phenolics, organic sulfur compounds, organo halogens, amines, quaternary ammonium salts, inorganic salts, formaldehyde, and organometallics. These chemicals act as toxicants to the microorganisms by rupturing and penetrating the cell walls and by interfering with oxidation of carbohydrates, and other mechanisms that stop the metabolic process.

Protein solutions and coatings are susceptible to attack by bacteria because protein is an easily utilized food source. The microbial attack causes a change in stability, viscosity, and pH. The anaerobic bacterial action also causes precipitation of protein, bad odor, and discoloration of the coating. Application of spoiled coating to paper may cause voids, spots, specks, and streaks in the coated paper. Uneven pickup of coating and poor adhesion may also result.

Each preservative problem must be studied individually to determine the proper chemical agent, method of addition, and quantity required to combat the microorganism present.

Flow Modifiers

Chemical flow modifiers are added to wet coatings to control, lower, and stabilize the viscosity of adhesives, and to regulate the flow of the finished coating mixture.

Urea, a highly soluble compound, lowers initial viscosities of starch and protein dispersions used in coatings. It is also effective in lowering the rate of increase of viscosity in starch coatings upon ageing. In casein coatings, it reduces viscosity and makes possible an increase in the solid content of the coating. Approximately 5% of urea on the adhesive solids is adequate to obtain satisfactory results.

Dicyandiamide lowers the viscosity of casein coatings without producing undesirable results. It is amphoteric and not as soluble as urea, being limited to 3.5% in cold water. It is useful in starch coatings to lower the initial viscosity and maintain lower viscosity ranges for long periods of time by preventing gelling. It is used in quantities of 0.5—25% on the weight of the adhesive.

Insolubilizers

Coated colored papers and coated paper for offset printing need a moderate degree of wet-rub resistance. Identification tags and coated poster board for use outdoors require a high degree of water resistance to withstand exterior exposures. Starch, protein, and synthetic adhesives vary in their inherent ability to resist water damage. Starch and protein coatings can be insolubilized by blending with resins or latices that are water resistant. The water-sensitive chemical groups of the adhesive can be reacted with a resin that blocks the tendency to go into solution. Another mechanism is to use cross-linking chemicals which cause the adhesive to polymerize and reduce the tendency to redisperse in water.

Insolubilization of Starch. Insolubilization of starch has been accomplished by reaction with formaldehyde, acid-catalyzed to yield water-insoluble compounds. Condensation products of formaldehyde with phenols, melamine, and urea develop suitable wet-rub resistance with starch upon ageing. One postulate on this mechanism is that the aldehyde and the hydroxyl groups of the starch form a complex insoluble compound. The cross shucking of starch molecules proceeds until the starch loses its ability to swell in water. Latices such as butadiene-styrene, polyvinyl acetate, vinyl acetate copolymers, and polyacrylics in combination with starch yield good water-resistant coatings.

Insolubilization of Protein. Formaldehyde is the most commonly used additive for insolubilizing protein. Formaldehyde reacts by replacement of an active hydrogen atom of the amino acid group in the protein molecule. The amount of formaldehyde that can react with the protein increases as the pH, temperature, and time of reaction are increased. Formaldehyde must be added in limited amounts and in dilute form to prevent excessive thickening of the coating. Formalin is usually diluted with equal parts of water and added slowly to the coating just prior to sending the coating to the machine. Casein coatings at a pH of 8.0 or lower, with casein cut by ammonium hydroxide, are reasonably tolerant to additions of formaldehyde. Coatings at low solids (40—50%) accept larger amounts of formaldehyde with little thickening.

Additions of formalin of 3—5% based on the protein increase the initial resistance to water and accelerate the rate of development of wet rub. Exposing the coating to elevated temperatures near 200-300° F also speeds the development of wet-rub resistance.

Paraformaldehyde is a mixture of polyoxymethylene glycols containing more than 90% formaldehyde. It is a solid that dissolves in hot or cold water and depolymerizes to free formaldehyde as it goes into solution. A major advantage in using paraformaldehyde is the slow release of formaldehyde to prevent increases of viscosity and to minimize buildup of fumes during handling.

Glyoxal is an active dialdehyde that readily reacts with protein and polyvinyl alcohols to form insoluble polymers by cross-linking.

Formaldehyde in combination with ammonia in water produces

hexamethylenetetramine (HMT). In pure form it is an odorless crystaline solid also known as hexamine. 1 lb of HMT is equivalent to 3.4 lb of formalin (37% formaldehyde). It reacts, and is used to obtain satisfactory results, in the same way as formaldehyde.

Dimethylol urea (DMU) produces insolubilizing effects on protein in paper coating in a manner similar to that of formaldehyde. It is added to coatings as a 10% aqueous solution in amounts up to 3% of active material based on the protein. Ammonium chloride is added to catalyze the hardening action on the adhesive.

Urea formaldehyde resin is composed of two parts formaldehyde and one part of urea. Melamine formaldehyde resin contains three parts of formaldehyde and one part of melamine. These resins are added to coatings in amounts up to 4% of the weight of the protein. Insolubilization occurs by reaction between the methylol groups and the active hydrogen atoms of the protein. Curing of the resin during drying also contributes to the wet-rub resistance.

Ultraviolet Absorbers

Ultraviolet radiation, or "black light," which represents only 5% of the radiant energy reaching the earth from the sun, is responsible for the photochemical degradation of many chemical products. This damaging radiation has wave lengths between 290 and 400 μ. Such energy, when absorbed, can change the color, physical form, and reactivity of many synthetic and natural structures. Typical manifestations of degradation following exposure to ultraviolet light include loss of dimensional stability in paper, fading of colors, crazing and embrittlement of coating, loss of tack in adhesives, loss of fluorescence in dyes and pigments, and hardening of flexible polymers. Substituted benzophenones, when added to coatings, can improve the stability of such materials by absorbing ultraviolet rays and harmlessly dissipating the energy. In a sense, these materials function as optical filters or shields.

COATING PREPARATION

All methods of mixing coatings deal with the processes in which a dry pigment is put into a slurry form. The state of dispersion of pigments in the vehicle is related to the fluidity of the final mixture. The degree of dispersion at the time of application of coating determines the perfection of coating laydown and affects the gloss and smoothness of the finished product.

The breakdown of aggregates or clusters of pigments is brought about by the application of forces that exceed the bonds between the pigment particles. Usually the force reaching the pigment clusters is dependent upon the power or total work put into the mixing effort. In low-solids suspensions, high energy input by high speed agitators may not cause adequate forces to bear on individual clusters to break them. Application of considerable force to pigment aggregates is also possible with modest power input. The *kollergang* or edge runner, is a mixer that concentrates power into a small volume. At moderately high solids, the colloid mill and impact mill also utilize this principle. Viscous media exert high disruptive internal shearing forces that create breakdown through interparticle friction. Water-base coatings at low viscosity can produce adequate disruptive shear only when violently agitated. Mechanical shear and chemical treatment are combined to disperse pigments. The chemical additives reduce viscosity and assist in overcoming the interparticle forces that cause pigments to aggregate.

Open-Rotor Mixer

The simplest (Figure 4-4) type of mechanical mixer is the open rotor which uses a vertical rotating shaft with blades or paddles attached at right angles. Shearing forces are generated between the moving agitator and the pigment, as well as within the turbulent layers of the mixture. Concentrated mixing action occurs near the surfaces of the rotating parts. The open rotor is commonly used for dispersing kaolin clay into freely flowing suspensions at 70% solids by weight. The vertical shaft can be equipped with bars, blades, solid discs, and paddles of various sizes and shapes. The mixing intensity can be increased by increasing the velocity of rotation, by increasing the concentration of pigment or by a combination of both.

Open-rotor mixers are more popular than other types because they can be used for a variety of applications. They can be easily constructed at low cost and are the oldest and best known. Regardless of the care in design, the open rotor is comparatively inefficient both as to power consumed and quality of results.

The diameter of the impeller relative to the size of the tank, and the speed, width, and angle of the blades are important factors in the design of an open-rotor mixer. Generally, the smaller the impeller, the higher the speed. Paddle-type blades are used in the speed range below 50 rpm. Large blades, which span the entire tank, are used to mix viscous materials where total movement is required to

OPEN ROTOR   KNEADER

ATTRITOR

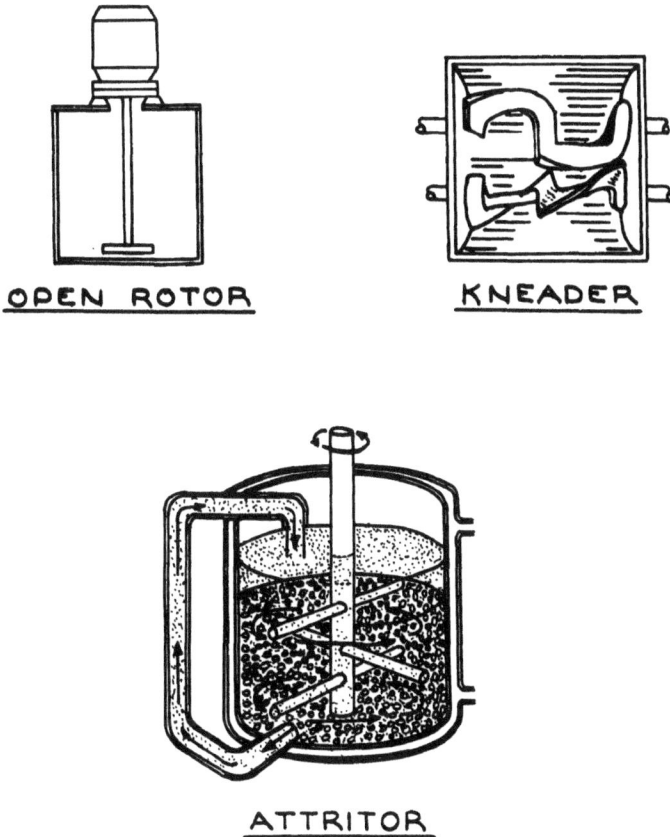

Fig. 4-4. Mixers [1].

assure that all the material in the mixer is subjected to positive mixing action. Small, thin, bladed arms are used for low-viscosity coatings to produce high shear and overall circulation at high speed.

The flow of coating within the mixer is determined by the design of vessel and impeller. Centrifugal force of the rotating impeller produces some radial flow. Coating may be drawn from the surface or directed against the bottom of the tank by inclining the impeller blades from the vertical. Rotation of the entire coating mass is to be avoided because it has no radial or vertical component of flow to aid in good mixing. High shear and maximum vertical movement can be realized by installing vertical baffles in the vessel. In low-viscosity coating systems, the width of the baffle is generally 1/10 of the diameter of the tank, viscous coatings require narrow baffles; and very viscous systems do not require baffles at all. Optimum results are obtained when the depth of the suspension is nearly equal to the diameter of the tank. The minimum depth for mixing is seldom less than one third of the diameter of the tank. Round or cone-shaped bottoms make possible efficient operation at low levels.

Semi-Enclosed Rotor Mixer

In another mixer, the rotor is partially or totally enclosed by a

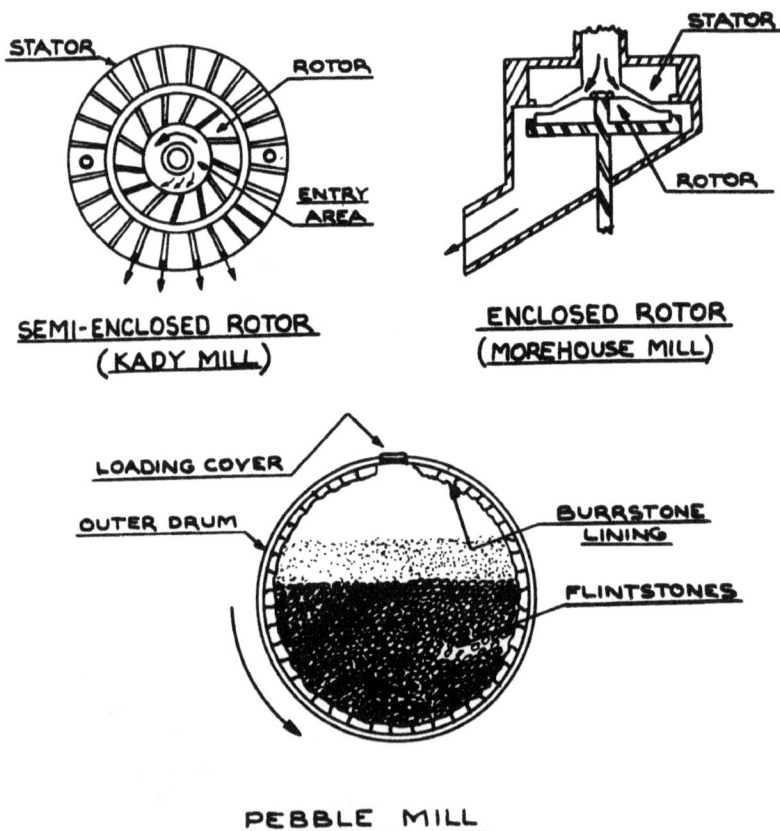

Fig. 4-5. Mixers [2].

Fig. 4-6. Kady mill installation (*Courtesy of Champion Paper Company.*)

stationary member. The Kady mill (Figures 4-5 and 4-6) uses a semienclosed rotor designed to force the coating rapidly through a slotted rotating member and through apertures in the stationary member, which encloses the rotor. The design utilizes the kinetic energy of particles in motion to increase the frequency of impingement or contact. The high power input creates high shearing force in the spaces between the rotor and stator. Impact of the fast-moving particles against the walls of the apertures in the stator provides intensive dispersing action. The rotor develops a velocity of nearly 7000 ft/min at the rim. Large amounts of energy are absorbed and utilized for dispersion work. A 12-in. rotor absorbs up to 150 hp in a vessel of 1000-gal capacity. This concentration of energy produces nearly perfect dispersion in a short period.

The Kady mill is capable of performing work on the coating mixture regardless of the consistency of the mix. As the viscosity increases, the amount of energy absorbed and converted to work of dispersion also increases. Frictional resistance is low because the rotor is small and the slots in the rotor through which the particles travel are short. The rotor is surrounded by a stationary member with numerous slots that act as valves on each of the slots in the high speed rotor. As the particle moves through the rotor slot, it is interrupted hundreds of times by the valving action. Each of these impacts contributes toward disaggregation and reduction in particle size. Instantly, the tremendous kinetic energy of motion in the pigment is 90% converted to work upon impact.

The finished coating can easily be made in one operation with this type of high-energy equipment. The vast input of energy results in generation of heat that is adequate for dissolving protein in the presence of alkali during the stage when the pigment is being dispersed. Specialty coatings that require ultrafine pigment particles and maximum color strength are produced efficiently and uniformly in a semienclosed rotor system. Dry colored pigments or pulp colors, emulsions, and plastisols can be easily processed to uniform control standards.

Enclosed Rotor Mixer

Colloid mills are based on the principle of the enclosed rotor (Figure 4-5). The coating, a fluid pigment, is passed through the close clearance between the rotor and the stationary member. Dispersion is accomplished by the internal shear in the fluid coating.

The energy input available for dispersion is controlled by the solids content, the viscosity of the coating, and the clearance between the rotor and stator. The continuously adjustable gap between the rotor and stator is the key to operation of a colloid mill. Precision design makes operation with clearance as low as 0.001 in. possible. The rotor may or may not be grooved and it may or may not be conical. The coating material is subjected to intense shear and centrifugal force, and the combination acts to exert considerable force for dispersion.

In operation, unrefined material is fed into the vessel where the upper turbine blades subject the coating to a whirling action at high velocity. The coating then passes through the adjustable clearance, which applies mechanical and hydraulic shear. Finally, the coating passes into an outlet pipe where it is either drawn off or fed back into the vessel for further refining. More efficient operation is obtained if the coating is first given a premix before actual passage through the shearing zone. An especially attractive feature of the colloid mill is the ability to recirculate and realize more than one pass through the mill. This mill is particularly useful in dispersing dry colored pigments or additives in white pigment coatings. The mixing is rapid and thorough and is particularly flexible for application in a specialty coating mill.

Kneader

The kneader (Figure 4-4), with two arms rotating in opposite directions in a vessel with a divided trough, is used for mixing thick, plastic coatings. Pug mills and internal mixers with sigma blades are also classified as kneaders. The parts move slowly, applying the force directly to the pigment through the high shearing stresses within the plastic mass. Dispersion is not dependent on kinetic energy as in the high speed mixers. The formulation and order of addition must be carefully regulated to obtain a coating that resists the motion of the moving blades. Percentage of solids, the coating rheology, the volume of coating relative to the blade, and capacity of the vessel influence the degree of dispersion.

Heavy sigma blades, helical in shape, rotate in opposite directions across the trough and effect lateral movement, kneading, and folding of the coating. The blades are sometimes serrated or toothed to increase the tearing action and to generate more shear. The blades overlap to produce a good transfer of coating from one blade and trough to the other. Nonoverlapping arms are operated at differential speeds with slight clearance so that they clean each other in passing.

High-solids coating up to 65% by weight requires approximately 30 hp/100 gal of capacity. The pigments can be dispersed first, followed by addition of the adhesive, or the pigment can be dispersed in the presence of the adhesive. Sometimes the pigment can be kneaded into the adhesive solution, thus keeping additions of free water at a minimum. Optimum operation is attained when the size of the batch is colse to the capacity of the kneader at the maximum percentage of solids. At high solids, the adhesive demand is reduced by forcing intimate contact between the adhesive and the surface of the pigment. This procedure enhances the adhesive-to-pigment retention and minimizes the probability of drainage of adhesive into the base paper. The kneader is a batch system and is flexible in batch size. Small batches cannot be processed efficiently.

Pebble Mill

The pebble mill (Figure 4-5) is a rotating cylinder in which flint pebbles, or porcelain or steel balls are used as grinding media and the inside of the steel cylinder is lined with porcelain, burrstone, or some other nonmetal liner. The metal cylinder operates at the optimum rotational speed for most efficient grinding. At optimum speed, the grinding medium forms a cascading, sliding stream of several layers of balls. The top layers travel faster than the lower layers and cause a grinding action. The gyration of the individual balls causes a rubbing or rolling contact between the balls. The sliding or cascading, is affected by the speed of the mill, amount of grinding medium, amount of coating in the mill, and viscosity of the coating. The critical speed is that speed at which the grinding medium, without coating, begins to centrifuge and resists cascading. The smaller the cylinder, the faster must be its angular velocity to attain critical speed. A 4-ft diameter commercial mill has an optimum speed near 20 rpm.

The mill should be at least half-filled with grinding medium for most efficient results. The wear can be reduced by filling the mill slightly beyond the half-way mark. Flint pebbles, and porcelain and steel balls are most commonly used. Flint pebbles obtained from the beaches of Normandy are exceptionally tough and long-wearing, outlasting most of the synthetic grinding media. Pure white ceramic porcelain balls are also used where high purity is required. Generally they do not grind as fast as flint pebbles, nor do they wear as well. The use of steel balls reduces milling time to one-third that required with other materials.

The correct size of the grinding medium is perhaps the most important factor for satisfactory operation and the smallest feasible should be used. The optimum size of the balls does not change with the size of the cylinder. Small balls provide more contacts per revolution, reduce the void size, and utilize the grinding energy more efficiently than large ones.

Approximately 30% of the volume of the cylinder is occupied by the grinding medium when the mill is half filled. The voids between the balls represent approximately 20% of the total volume. The coating should fill the voids and just cover the balls for fastest grinding. Excessive wear occurs when the coating charge is allowed to drop below the surface of the grinding medium.

The viscosity of the coating is optimum when it is equivalent to that of a freely flowing pigment slurry. Low viscosity allows the balls to strike against each other with too much force and causes excessive wear. When the viscosity is too high, the balls are not free to cascade and roll inside the mill,. No grinding occurs when the balls are carried with the viscous coating.

Attritor

The attritor (Figure 4-7; see also Fig.4-4) resembles the pebble mill in grinding action and differs primarily in that the grinding medium and the coating charge are ground inside a stationary vertical cylinder by an agitator. Two series of agitated mills have been developed–one for batch work, the other for continuous operation. The effectiveness of the ball mill principle is determined by the number of contacts occurring between the grinding medium. The smaller the balls, the greater the number of balls per unit volume and the larger the number of possible contacts. Decreasing the diameter of the ball by one half increases the number of contacts four times and causes a corresponding increase in grinding rate. As the ball size decreases, so the weight decreases, reducing the impact force between balls and leading to complete suspension of the grinding medium in the coating. Therefore, utilization of the small balls requires agitation by additional mechanical action.

The attritor functions through an internal agitator that is operated by a drive motor located above the vessel. The attritor, with small balls agitated from the outside, increases the grinding rate and provides exceptionally fine grinds.

The cylinder is constructed from stainless steel. An 80-gal working

Fig. 4-7. Attritor coating preparation (*Courtesy of Appleton Coated Paper Company*).

capacity requires about 900 lb of flint stones or 2,400 lb of steel balls. A 10-hp motor drives the unit and continuous recirculation is desirable. The kinds of grinding medium available are the same as for the pebble mill. The sizes vary from 1/2—1/32 in. in diameter. Depending upon the formulation, the attritor is capable of grinding at rates 10-100 times those of the pebble mill.

COATING METHODS

In a general way, modern coating installations are divided into two groups. In the first, the coating film is applied after the base stock has been manufactured, the coating equipment being entirely independent of the paper machine. In the second, the coating is applied while the paper is being produced, the coating equipment being an integral part of the paper machine. The terms "off the machine" and "on the machine" are descriptive of these two general classes. The "on-the-machine" method, known also as "machine coating," has gained wide

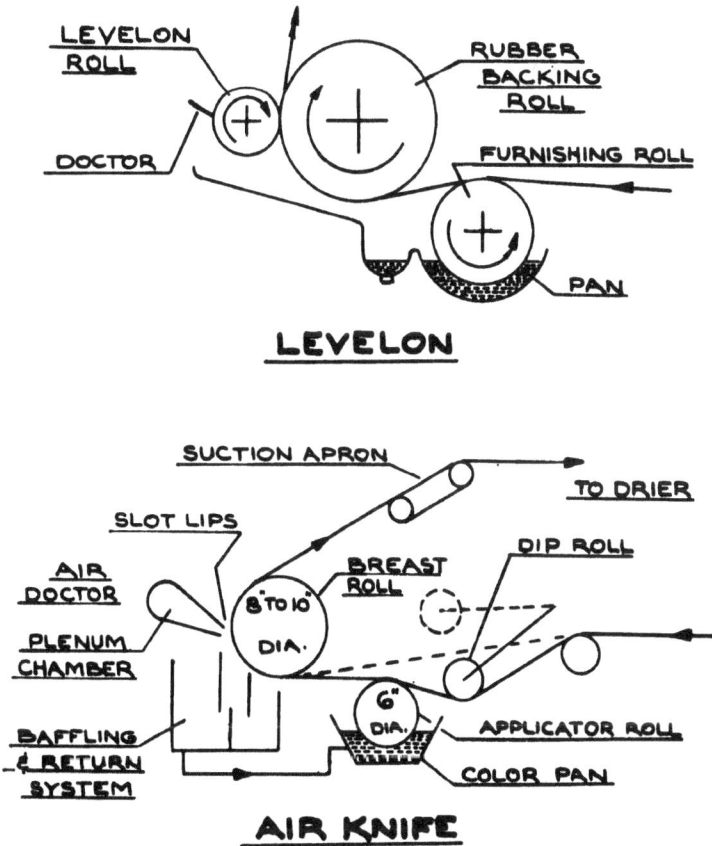

Fig. 4-8.   Off-machine coating methods [1].

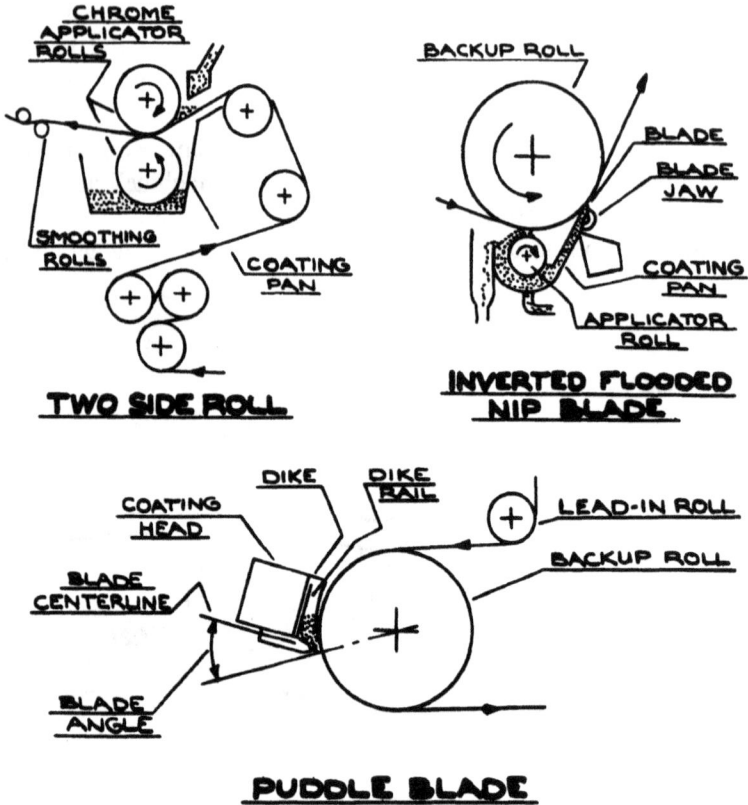

Fig. 4-9. Off-machine coating methods [2].

acceptance and has been motivated by a need for large tonnages of magazine, catalog, and label papers, economically produced and superior to filled supercalendered papers.

All coating methods, regardless of mechanical design, incorporate these basic elements: (1) application or transfer of coating to the substrate, (2) metering to regulate the coating weight, and (3) smoothing or leveling to obtain a plane surface. The various coating machines are engineered to achieve these features either sequentially or simultaneously. Rolls of various sizes and materials are arranged in many ways to transfer, meter, and smooth coatings. Rolls in combination with air-knife metering and smoothing rolls are employed. Finally, the trailing blade, alone or in combination with rolls, performs the necessary functions to accomplish the elements of coating. The process has been incorporated as an integral part of the paper machine and also as an entirely separate "off-the-machine" method (see Figure 4-8 and 4-9). The methods described refer to the latter type of installations.

Air-Knife Coater

The air-knife coater (Figures 4-8, 4-10 to 4-13) is generally used as an off-machine coater to apply coatings at low solids and viscosity.

Fig. 4-10. Modern high-speed air knife coater (*Courtesy of Appleton Coated Paper Company*).

Fig. 4-11. High-speed air knife coater (*Courtesy of Appleton Coated Paper Company*).

The nature of the process at low solids requires some drying before the paper can contact rolls or drying cylinders. This factor has traditionally restricted the use of the air-knife coater to an off-machine process except on paperboard machines at low speed. However, recent advances in high-efficiency driers have contributed toward minimizing this limitation.

Air-knife metering provides the advantage of uniform distribution of coating over all areas of the sheet. It is relatively simple to operate and is extremely versatile. The method is easily applicable to a wide range of rawstock weights, widths, and different grades of coating.

Applicator Roll

The applicator roll, or series of rolls, transfers the coating from the color pan to one side of the sheet. The transfer roll is almost always moving in the same direction as the paper, resulting in a film split that produces undesirable ridges. Sometimes this condition is rectified by installing a smoothing roll between the applicator and the air knife. The coating layer is metered by the flexible air jet, which cuts off the excess coating and leaves a uniform film on the paper. Generally, large excesses are applied ahead of the air knife to assure a uniform, skip-free surface following the metering action of the air knife. The applicator roll acts as a pump and supplies quantities of coating directly related to the roll speed.

The air-knife chamber is a reservoir for air under pressure, which is forced out at high velocity through a narrow slot. The slot is formed with carefully machined lips that can be adjusted to regulate the thickness of the air stream. The geometry of the air-knife relative to the backing roll must be controlled at optimum conditions for satisfactory performance. Orifice opening (0.25-0.35 in.), the gap between the knife lip and the paper on the backing roll (0.050-0.140 in.), and the angle of the air stream vary according to the grades produced.

Fig. 4-12. Modern off-machine high-speed air knife coater [1] (*Courtesy of Waldron Corporation*).

Fig. 4-13. Modern off-machine high-speed air knife coater [2] (*Courtesy of Waldron Corporation*).

Variables

The amount of excess coating blown off by the air stream is regulatèd by changes in air pressure, usually 1—6 lb/in². The coating viscosity and the web speed influence the pressure required for adequate metering. As the viscosity and speed increase, greater air pressure is needed. The operating speed varies between 400 and 1500 ft/min depending on the efficiency of the drier and the grade being processed. Normally, coating solids range between 20 and 55% with 40% as an average. A normal operating viscosity would be between 100 and 300 cP for Newtonian rheology. A wide range of coating weights, from 1 to 40 lb/3000ft², is easily attained.

Puddle-Blade Coater

The puddle type of trailing-blade coater (Figure 4-9) applies, meters, and levels the coating in one simultaneous action. The apparatus consists of a flexible blade held firmly in metal jaws, dikes at the ends of the trough, and a back-up roll. The paper carried by the back-up roll forms one side of the pond. The trough and blade are forced against the backing roll, causing the flexible blade to deflect downward. The blade is made from spring steel and is usually 0.012 in. thick. The blade extension, angle of contact, and point of contact vary according to the particular application. A representative extension of the blade beyond the jaws is 3/4 in. and the point of contact is near 8:30 o'clock on the back-up roll. The angle made by the blade and the radius of the back-up roll at point of contact is generally near 30°. The composition of the back-up roll varies from rubber covering to solid metal.

Whatever the material of construction, it is necessary to keep the surface free from grooves, scratches, burrs, and ridges. Most operators carefully screen the original and recirculated coating to remove particles and fibers, which could cause serious scratch defects in the coating.

The end dikes are curved to fit the surface of the back-up roll with minimum clearance to prevent the coating from leaking out. Felts and fluid seals are sometimes used with coatings of lower viscosity to improve the seal on the dikes.

Inverted-Blade Coater

The inverted-blade type of flooded nip coater applies coating to paper with a transfer roll and simultaneously meters and levels by means of the flexible blade. The paper web wraps the back-up roll, an excess of coating is applied by a rotating roll, and the excess is removed with a flexible blade about 0.012 in. thick.

Variations in the application step include multiple-roll systems of various sizes and arrangements. The fundamental principle is similar, in that controlled transfer is accomplished by a rotating roll or rolls. This arrangement offers more control over uncoated edges, has no leaking-dike problems, and the coating supply can be removed from the sheet without discharging the coating from the station. Recovery and ease of clean-up following a break are also less troublesome with the inverted-blade arrangement. A wider range of coating weights is possible with the inverted blade. Control of coating weight involves regulation of blade pressure, blade angle, applicator roll gap, roll speed, coating solids, and paper speed.

Trailing-Blade Coater

The trailing blade principle with many mechanical variations is used for publication papers on and off the paper machine. Arrangements involve coating one or both sides either alone or in combination with roll coaters to apply two layers to each side. Conversion coaters for coating both sides of publication papers operate regularly at speeds near 3000 ft/min. The maximum speed of the coater is determined by paper strength and safe drying capacity rather than by the coating method itself. Web handling problems concerned with automatic splicing and control of tension must be overcome. Coating solids between 50 and 65% give coating weights between 6 and 12 lb/3000ft² on papers of publication weight. The coating viscosity is normally high, near 15,000 cP at 12 rpm on a Brookfield viscometer. Higher viscosity tends to increase the hydrodynamic pressure and decrease the blade pressure, thus allowing higher coating weight.

The main advantage of the trailing-blade coated papers is the levelness of the surface at any coating weight. The blade coater produces a level surface at the expense of nonuniform thickness of coating layer because of the tendency to fill in the valleys of the paper surface and to skim the peaks. For white publication or printing papers, where the coating and base stock are similar in brightness nonuniformity causes no problem. For tinted coatings or where great contrast exists between coating and base stock, an undesirable mottle or pattern results. For functional coatings of low coating weight, requiring uniform thickness from point to point on the surface, the blade may not be satisfactory.

Levelon Coater

The Levelon coater (Figure 4-8) can be likened to an air-knife coater except that the metering and smoothing are done by a reverse rotating roll rather than by a jet of high-speed air. It is capable of applying water-base coatings of high solids at high speeds over a wide range of coating weights. An excess of coating is applied to the web by one of a variety of furnishing arrangements. The pressure-loaded nip between the Levelon roll and the rubber backing roll meters and levels the fluid coating. An oscillating doctor blade assures a continuous clean surface in contact with the coating.

Coating solids between 35 and 65% yield coating weights of 3—40 lb/3000 ft². The coating weight is controlled by variations in nip pressure for a given viscosity, line speed, and rubber-roll hardness. The speed and surface perfection of the Levelon roll controls the surface characteristics of the finished product.

Two-Side Coater

The flooded-nip two-side coater applies coating simultaneously to both sides of the web of paper. Coating is applied to the bottom side by a conventional roll transfer and to the top side by flooding the sheet ahead of the nip. Metering is accomplished by adjustment of the nip opening. Smoothing or leveling is done by successive reverse rotating shafts. Two-side coaters have proved to be versatile machines capable of applying coatings of 20—45 % solids and achieving coating weights of 2—20 lb per side (24x36-500). Speeds up to 650 ft/min have been attained with appropriate drying. The paper enters the applicator nip tangentially to both rolls; the nip opening is controlled by adjustment of the top roll only. Usually, a clearance of 0.001 in. above that of the paper caliper is equivalent to 10 lb of coating per ream (24x36-500). The speed of the lower roll is synchronized with that of the paper web. The top applicator roll may be run slightly faster than the web to allow for the sheet to follow or wrap the top roll slightly as it leaves the nip. Most of the coating film is transferred to the paper under ideal conditions. However, some film splitting occurs, leaving on the coating surface a pattern that is rectified by the smoothing rolls. The point of film split and degree of pattern are affected by thickness of the film, coating solids, breakdown of viscosity with shear, absorbency of the base stock, and the water retention of the coating. Each of these factors must be controlled for optimum performance.

DRYING COATED PAPERS

Drying of coated paper involves two fundamental processes: transfer of heat to evaporate water and mass transfer of evaporated liquid in the form of vapor. Production driers may accomplish heat transfer by convection, conduction, radiation, or a combination of these mechanisms. The method of heat transfer is the basis for classifying commercial driers. Normally, the heat flows from the outer surface of the coated sheet to the interior of the layer. Drying by high frequency electricity, which generates heat internally, is a method that results in heat flow from the interior to the external surface.

Festoon Driers

The oldest and best-known air-convection driers are those that employ the straight-line and return-line festoon system. The wet coated paper is formed into loops on rods or sticks fastened at their ends to chains that carry them through the drying chamber. Hot air from 80-150°F is supplied through ducts at the top and removed through exhausts at the bottom. Drying paper without tension and minimum change in strength or color is possible at low drying rates in festoon driers. For expensive coated papers, open festoons allow for easy and total inspection. Simple and inexpensive web-handling equipment can be used with festoon driers.

Uneven drying from top to bottom of the loops, slow-speed pick-up and take-down of the loop, and distortion of paper where the web hangs on the supporting sticks are among the disadvantages.

Tunnel Driers

The straight-line tunnel drier conveys the coated web at speeds up to 1500 on conveyor bars through a long insulated oven. Hot-air ducts equipped with discharge holes or nozzles distribute hot air across the full width of the web. An air temperature of 350°F is obtained with steam heat and temperatures up to 500°F are possible with direct-fired gas burners. Evaporation rates of 1—3 lb of water /(ft^2)(hr) make high-speed and moderate-length driers practical.

Drier Zones

The tunnel drier is usually divided lengthwise into sections or zones to provide variations in temperature and air velocity in different parts to regulate the drying rate, as the product demands such control. Air velocities at the point of impingement are usually between 2000 and 3000 ft/min. Higher velocities tend to produce flutter and depress the web between support bars, which are 12—18 in. apart. The mechanics of tunnel driers design imposes practical limits on speed and width. Wide driers require conveyor bars of large diameter to prevent sag and deflection. The diameter of bar or stick seldom exceeds 2 in. The bar is still subject to deflection and occasional breakage when centrifugal forces act at the turning sprockets.

Impingement of Air

Carrier conveyors of the arched-belt type offer continuous support to the entire web and permit improved web handling. Air velocities up to 12,000 ft/min can be used to increase evaporation rates and increase production speeds with shorter driers. The conveyor and web can be run at identical speeds to eliminate excessive tension during the drying cycle.

Bar marks and deflection do not impose mechanical limitations on the speed of operation. Impingement of air at high velocity on the surface of the paper is possible without supporting the sheet with a large-diameter drying drum or series of drums. Evaporation rates of 6—8 lb of water/(ft^2)(hr) are possible without damage to the coated product.

Air "Floater" Driers

The air "floater" drier is required for coating equipment that simultaneously coats both sides of the web. The floater is generally of sufficient length to assure that the underside of the web is dry enough to contact a transfer apron without damage to the surface. Following the transfer apron, the drying is completed in a festoon or tunnel drier similar to those used for one-side work.

A column or pillar of hot air is introduced to the bottom of the web to dry the underside and float the sheet under tension in the space between the coater and transfer apron. The amount of drying accomplished in the floater on the bottom side of the web regulates the speed of the machine. This in turn is determined by the volume and velocity of the bottom air that can be tolerated without excessive lifting, flutter, or sheet instability in the floater section.

Cylinder Driers

Cylinder driers involve heat transfer by conduction from the internal heat source through the thick metal wall to the web of paper in contact with the hot surface. Such cylinders are arranged in series in two horizontal rows, one above the other, with the upper cylinders above the spaces between the lower cylinders. High-velocity air hoods or caps are often constructed over the steam-heated cylinders. The discharge nozzles have slot openings of 0.035 in. and a gap of 0.25—0.75 in. between the nozzle and web. Evaporation rates up to 35 lb of water/ $(ft^2)(hr)$ are attainable with such driers.

Radiant Driers

Radiant or infrared driers are frequently used alone or in combination with air driers. The heat is generated by electric incandescent lamps or gas-heated incandescent refractories. Infrared driers, utilizing lamps or refractories, are arranged as a tunnel through which a conveyor carries the coated paper. The temperature at a given line voltage causes electric units to transmit infrared rays at a definite wave length. Gas units can vary the temperature of the particular refractory and transmit a wide band of useful infrared energy. Penetration of the radiant energy into the coated web depends on the wave-length involved and on the absorption properties of the particular coating and base stock.

To obtain optimum efficiency, the major band of wave lengths of the transmitting source must be regulated to transmit a majority of the wave-length that is effectively absorbed. The remaining energy is reflected and does not contribute to effective drying. Infrared heating sources are of particular value for supplying rapid drying without causing surface hardening or "skimming." Hot air is used in combination with infrared rays to facilitate removal of vapor from the surface of the web.

Dielectric Driers

Dielectric drying can generste heat in the coated web by placing it in a strong electrostatic field produced by high-frequency voltage. A dielectric drier in its simplest form consists of two flat metallic plates, between which runs the coated paper. Electrically, this represents a capacitor, the plates being connected to a high-frequency electronic generator. As the alternating current rapidly reverses the polarity of the plates, the molecules rapidly oscillate, thereby generating heat within the coating layer.

The outstanding characteristic of this form of drying is the uniformity of heating throughout the thickness of the coating layer. The responce of wet coated paper to dielectric drying is such that the potential for heat generation is greatest on the wet paper. As it dries, the responsiveness decreases in a regular way. This principal advantage brought about specifically by dielectric heating provides the opportunity to produce an even moisture profile.

Many factors must be carefully considered in the selection of the most suitable drier for each coating application. Driers are not usually flexible enough to adjust for major design errors or their influence on product quality. Therefore, experimental tests should be conducted on the products in question before a system is finally selected. The properties of the coated paper being processed, the drying characteristics of the coating, the quality of the final product, the utilities and facilities available in the plant, and data on cost and performance are all important in choosing or designing the best drier most economically.

CALENDERING

The finishing operations in a coating mill begin immediately after the coated paper leaves the reel of the coating machine. They consist of calendering and cutting the paper into sheets or winding into well-formed rolls, inspecting and removing defective sheets, counting, trimming, sealing, labeling, packing, and shipping. The finishing department, because of the diversity of operations, is of greater importance than it was in the past. Paper arriving in the finishing department must be free from faults to keep waste at a minimum.

Calenders

Calenders and supercalenders are similar in general appearance and construction, each composed of a series of rolls mounted vertically in upright side frames and equipped with a means of applying pressure as the paper passes between the nips of the rolls. A calender is composed wholly of metal rolls, whereas the supercalender has a combination of metal rolls and softer rolls made of paper or cotton fibers. The difference in types of roll accounts for the differences in gloss and smoothness obtained from the two operations.

The cast-iron chilled rolls in a calender retain their contour and are nonresilient at the nip contact. The gloss and leveling action in a metal-to-metal nip originates from the plastic properties of the paper. These properties allow the paper to compress and change caliper. A calender contains 3 to 10 metal rolls, the number and diameter being dependent on the kind and width of paper to be calendered. Pressure on the nip in addition to the weight of the rolls is applied by a compound lever system, power-screw method, pneumatic or hydraulic cylinders, and a compounded top lever.

Supercalenders

On a supercalender (Figure 4-14), the inert metal roll causes a de-

Fig. 4-14. Supercalender (*Courtesy of Appleton Coated Paper Company*).

pression in the filled cotton or paper roll at the point of nip contact under pressure. As the rolls rotate, the soft roll flows or creeps because of the tendency to return to the normal nondeformed state. This creeping causes a motion of the surface of the filled roll on the hard smooth surface of the metal roll, producing a polishing or friction action to provide high gloss and smoothness. The intensity of the friction is a function of the deformation in the rolls, the quantity of paper passing through, and the nip pressure.

The rolls are arranged similarly to those of a calender and vary in number from 3 to 9 for a single finisher and from 4 to 12 for a double finisher. The single finisher for coated one-side paper consists of alternate metal and filled rolls; the double finisher for coated two-sided paper has one extra filled roll to produce a nip with two adjacent filled rolls. This arrangement is necessary to ensure equal finishing on each side.

The metal rolls are made of cast iron treated to chill the outside of the roll to a depth of 1 in. to produce a hard surface. This surface can be polished to high finish and is durable. The filled rolls are constructed from cotton or paper fibers. The fiber material in the form of sheets is cut into disks of octagonal or circular shape. The disks are assembled on a shaft subjected to pressure and held fast under compression by metal ends screwed onto the shaft. The rolls are given a satisfactory finish by turning and polishing in a lathe to a uniform, smooth surface.

Gloss Calenders

A gloss calender consists of a highly polished drum of chrome-plated or superfinished cast iron heated internally to 300°F or higher. Pressure rolls made of rubber or plastic form a nip with the metal surface at a pressure of 350—500 lb/lineal in. Gloss calendering is done on coated paperboard grades and on some coated book papers. Because of the higher temperature, higher gloss and smoothness can be obtained with lower nip pressures, thereby retaining bulk, strength, brightness, and opacity.

APPLICATIONS OF PRINTING PAPER

High quality printing by all methods involving ink transfer requires uniform and total contact between the printing member and the paper surface. This requirement is especially important in halftone reproduction, where the image to be transferred to the paper is composed of individual dots, sometimes as many as 22,500 per square inch. A properly formulated pigment coating, applied and calendered to optimum levelness, provides such a planographic surface.

Letterpress Printing

Letterpress, or relief, printing remains an important method with respect to usage of paper. One of the most important requirements is printing smoothness—a combination of surface levelness and compressibility. The minerals and adhesives in a coating layer are generally more dense and less resilient than in uncoated papers. Therefore, a smoother coated surface is required. Elastic adhesives such as synthetic latex polymers contribute greatly toward coating resiliency. Precise balance between smoothness and resilience provides for minimum pressures, which yield optimum quality.

The use of letterpress inks, which are not as tacky as offset inks, results in less tensile force normal to the coated surface during the printing operation. This circumstance permits lower requirement of adhesive to meet the minimum bond or pick resistance, all of which favors the production of a smoother surface.

Uniform ink absorbency, cleanliness, whiteness, and opacity are also required. Hence, a formulation must be constructed with the proper kinds and quantities of pigment and adhesive to give adequate strength, balanced with levelness, resilience, and proper receptivity to ink. Freedom from loose surface particles, which can be trapped to cause repetitive defects in printing, and avoidance of abrasive materials, which tend to wear the metal plates, are also essential. An alkaline surface for letterpress printing is required to assure proper drying of ink by oxidation. Acids form a destructive chemical reaction to inactivate ink dryers.

A typical letterpress formulation is as follows: 50 parts No. 1 domestic clay, 25 parts fine fraction English clay, 25 parts precipitated calcium carbonate, 1/4 part sodium hexametaphosphate, 10 parts oxidized starch, 5 parts synthetic latex polymer.

Presses that use paper in web form often use inks that dry by solvent evaporation in high-temerature ovens. The coating must withstand the high temperatures and be sufficiently permeable to allow water vapor to escape from the sheet without blistering. Proper balance between moisture, particle packing in the coating layer, and discontinuity of adhesive is required to satisfy such demands.

Flexographic Printing

The flexographic printing process, also a relief method, uses rubber printing plates and low viscosity inks (water or solvent-base) that dry by evaporation in high temperature ovens. The paper is always fed into the press in web form. The flexible rubber printing plate can conform somewhat to irregularities in the coated surface. Controlled absorbency to assure good ink transfer and anchoring is required without feathering or spreading of the image. Uniform point-to-point absorbency is essential to assure a mottle-free printing surface. The fluid inks possess little tack and greatly minimize the adhesion and cohesion requirements of the coating layer. The surface chemistry is less critical, except as concerns ink drying. Extremes of alkalinity or acidity may have an effect on the stability of colored dyes or pigments used in the inks. A minimum resistance to water is required when water-base inks are used.

Offset Printing

Offset printing is done from a planographic surface rather than by means of a relief process as in letterpress and flexographic. Sensitized metal or paper printing plates carry the image, which is ink-receptive. The nonimage areas are water-receptive and therefore repel the ink. During printing, the entire surface is exposed to a water dampener, followed by the ink roller, leaving ink only on the image area. The ink image is transferred or offset onto a rubber cylinder and then to the paper. The thin ink layer transferred to the paper is viscous and tacky, a condition that produces a strong tensile force perpendicular to the surface and necessitates a strong coating bond and internal fiber bonding in the body stock.

Nature of Coatings

The application of moisture to the nonprinting areas requires that the coating be water resistant, lest the surface becomes weakened and picks off in subsequent impressions. Small particles, such as slitter or trimmer dust, fibers, and coating flakes, tend to adhere to the rubber blanket or printing plate and produce random defects on all succeeding sheets. The demand made on the paper may vary with the specific offset job involved—large solid areas in the printed matter create more stress than line or type work.

Multicolor work involves inks with decreasing tack for successive colors to ensure good ink trapping in overlapping sections. Consequently, a multiple-station offset press would start with a tacky ink and gradually decrease tack for subsequent layers, whereas an ink of lower tack may suffice for a one-color press. This sequence, coupled with the repeating water treatment, places more rigid requirements on offset coatings for multiple color work. Caution must be exercised to prevent possible extraction of surface-active material from the coating by the press fountain solutions. Soluble coating defoamers, free alkali, eveners, or adhesives may be transferred to the ink and produce an emulsion of ink in water. This emulsion can adhere to the nonimage area of the printing plate and appear on the paper to produce undesirable scumming. A typical formulation for offset coating follows: 50 parts No. 1 domestic clay, 40 parts fine fraction English clay, 10 parts precipitated calcium carbonate, 1/4 part sodium hexametaphosphate, 10 parts casein, 7 parts latex polymer.

Offset Paper

Recently, numerous multicolor presses for heat-set offset work have developed a large demand for light-weight coated offset paper in web form. The high-speed press creates a greater tension at the coated surface. Fortunately, this pulling force is partially neutralized because the heat-set inks are less tacky than regular offset inks. By printing the paper in web form with the grain, one can take advantage of the greater tensile strength in the machine direction and increased resistance to coating pick. Low moisture content and high coating porosity help to avoid blistering in the high-temperature heat-set ovens.

Smoothness requirements are similar to those for flexographic printing because the flexibility of the offset blanket allows it to conform, within limits, to the surface of the coated paper. Again, uniform ink absorption assures a mottle-free print. Properly controlled absorption to obtain good holdout for gloss printing must be balanced with sufficient penetration to avoid transfer of ink to the back of the succeeding sheets in a stack of freshly printed paper.

Dry offset printing, commercially termed "letterset" is a process in which the ink is printed onto the offset blanket from the raised image surfaces of a printing plate. The demands made on the paper by this process are similar to those for conventional offset lithography except that resistance to wet pick is not required.

Gravure Printing

The gravure printing process requires a metal printing plate or roll with the image etched in the surface in the form of cells with controlled shape, area, depth, and wall size. The ink must be adequately and uniformly transferred from the etched cells to the paper surface. The excellent uniformity and tonal gradation possible by this method are largely dependent on satisfactory surface properties of the smooth, properly composed, coated surface. The inks used for gravure printing are nontacky, solvent-based fluids that exert little pulling force on the coated surface. Consequently, this process does not need the high pick resistance required for offset printing. Absorptivity of the coated surface and proper cell-to-paper contact governs the quantity of ink transferred and controls the quality of the final image.

The coating formulations generally include soft compressible materials that can be calendered with relative ease. Accurate reproduction of rotogravure highlights free from harsh grainy appearance is realized only with uniformly controlled smoothness and compressibility. Recent advances in the technology of adhesives and chemical additives have contributed much to this realization, and the future shows considerable promise for greater improvements.

Screen Printing

The screen printing process requires a fine mesh screen, usually silk, through which heavy layers of ink are transferred directly to the paper in the image area. Slow and costly, this process is limited to special applications such as posters, show cards, and unusual package displays. The paper printed by silk screen must have good oil holdout, because the heavy layers of fluid ink remain in contact with the surface for long periods. The ability of the screen to conform to the substrate surface and the thick fluid layers of ink make less demand for smooth or level surfaces.

Electrostatic Printing

Major investments have been made in coating equipment to develop and produce smooth compressible surfaces for conventional printing. Recently, a new electrostatic printing process that can print on any surface, rough or smooth, has been introduced. This new method eliminates the need for pressure and uniformity of contact at the printing surface—no direct contact of the printing member with the material to be printed is made. This revolutionary technique is based on the use of an electrically-charged conducting stencil that serves as an image-forming master. The stencil is a conductive metal gauze or nylon, 133 lines to the inch for excellent quality and 250 lines for high-resolution work. The open-screen portions of the stencil form the active printing area, whereas the nonprinting areas are blocked off by a suitable polymeric substance.

A conductive backing plate or ground serves as a sheet electrode, which is placed parallel to the screen, leaving a space between the two members. An electroscopic ink in the form of finely divided powder is applied to the screen. These particles, electrically charged, move through the image areas of the screen toward the oppositely-charged conductive back plate. Paper or any other material interposed between the screen and plate is printed with a faithful reproduction of the image on the screen. The ink particles travel in a straight line toward the back plate until stopped by the intervening paper or object. Regardless of the surface contour of the article, the particles are deposited on high and low areas alike. Heat or solvent vapor is used to fix the image.

The ink technology is different, because of the finely-divided resin powders with special electrical properties. The particles are usually freely flowing spheres 1—5 μ in diameter. As in silk screen printing, multiple colors can be applied to a single area by using a different screen for each color, one following the other.

This method is satisfactory for surfaces that are otherwise difficult

Fig. 4-15. Jagenberg cutter (*Courtesy of Appleton Coated Paper Company*).

Fig. 4-16. Vacuum sealing (*Courtesy of Appleton Coated Paper Company*).

or impossible to print upon. Printing on apples, loosely woven fabrics such as burlap and terry cloth, corrugated paper board, odd-shaped plastic containers, and other extremely rough surfaces has been successfully done by means of the electrostatic printing method.

Overprinting

Solutions of film formers in organic solvents and clear nonpigmented printing inks are frequently applied over printed paper to improve gloss and resistance to scuffing and soil. The coating should resist penetration of solvents, the most common being alcohol, the vehicle in "spirit varnishes." Excessive penetration produces little gloss and loss of continuity for surface protection.

Satisfactory varnish holdout is attainable by maintaining adequate coating weight, sufficient and proper binder to provide a continuous film barrier to solvent penetration, and close pigment packing with fine particle size, low-absorption coating clays.

Figures 4-15 and 4-16 show typical cutting and sealing operations.

SPECIAL FUNCTIONAL PAPERS

Paper Offset Masters

Direct-imaged offset masters, not concerning photographic or sensitized types, are special coated papers, the surface coating performing as a lithographic printing plate on an offset type duplicating machine. Compared with metal plates, paper masters are relatively inexpensive to produce and simple to prepare for use. The trade requirements for paper offset plates are manifold. They must be:

1. Low in cost, light weight, and produce a large number of copies with the appearance of the original;
2. Suitable for use with distilled water and a variety of fountain solutions and etches;
3. Able to ensure good reproduction with ordinary typing-ribbon, pen-and-pencil, and Xerox images.

A paper coating 0.001—0.005 in. thick must have the property of rendering the paper not only water receptive, but ink repellent when wetted. This property makes the paper suitable for offset printing. To be satisfactory, the surface must also be ink receptive or oleophyllic to a degree where it receives and holds the image-forming material and printing ink. The surface must also be sufficiently hydrophyllic so that it receives and holds enough aqueous ink-repellent fountain solution to limit the reinking of the plate to the image area only.

If the water absorption is too high, the wetting solution undermines and invades the image, with the result that the image is lost or "walks off" the plate. Excessive absorption might also distort the paper substrate. On the other hand, if the surface is not adequately absorptive and fails to take on a sufficient quantity of wetting solution in the nonprinting areas, ink is deposited in these background portions, producing undesirable specks or scumming. Usually, the coatings that retain fountain fluid do not sustain erasures without destroying this oleophyllic–hydrophyllic relationship.

During actual use, the paper master is wetted from a fountain solution by rolls and the portions in the nonimage area are wetted uniformly. The image portion remains dry. The plate is then inked from a fountain by means of rolls where only the oleophyllic image accepts the ink. The ink is then transferred to the offset rubber blanket, which prints a replica of the image on the paper.

A typical coated paper master might be prepared in the following manner. On a suitable web of paper, sulfite, or kraft of a high wet strength, one or more coatings are applied to one side to seal the fibers. This waterproof coating must be a uniform, insoluble layer that provides good bond to the fiber and to subsequent coatings. The base coating must be insoluble to prevent picking or delamination, which is caused by water penetrating into the base stock. The base coating should also possess some lithographic ability to act as a secondary printing surface when erasures abrade through the top layer of coating. The coating could be casein, urea–formaldehyde, melamine resins, cellulose derivatives, chlorinated rubber latices, or combinations thereof. A surface coating suitable for printing is then applied.

This coating usually consists of an adhesive and an inorganic filler pigment such as barium sulfate, kaolin clays, calcium carbonate, and titanium dioxide. Metal salts such as aluminum chloride or zinc chloride are used to insolubilize the colloidal adhesive. Usually the process requires four steps; waterproofing, printing, coating, and the solubilizing wash coating. A water-insensitive coating is also applied to the back side to stabilize the sheet against curling.

Battery Separator Papers

Coated papers are placed between the negative zinc electrode and the depolarizing mix of a dry cell to improve operation. In cells of the Leclanché type, the zinc of the negative electrode is consumed by electrochemical corrosion, which sets up the chemical reaction necessary to produce the flow of electrons, thus creating an electric current. One important feature of the operation of the cell is the control of such corrosion that it proceeds uniformly over the area of the metal. Uneven corrosion results in decreased output capacity and may cause local puncturing of the zinc electrode. Puncturing usually leads to electrolytic leakage and a premature end to the life of the cell.

Purpose of Paper Separators

The purpose of a surface-coated paper separator is to promote uniform corrosion of the zinc electrode and thereby gain maximum utilization of the zinc with high output capacity and optimum cell life. This paper, coated with a gelatinous material, contains a mercury salt with limited solubility in water. Mercury salts of substantial solubility in water have been used in the cell electrolyte to reduce zinc corrosion on open circuit. Tests show that mercury salts of limited water solubility, when used in a paper separator between the mix and the zinc electrode, can reduce corrosion on open circuit and also produce a more uniform consumption of zinc on closed circuit.

Flashlight Cell

A typical flashlight cell is constructed with an open-top zinc cup, which constitutes the negative electrode. The bottom of the cup is covered with a layer of nonconductive material, which may be a disk of paperboard or pulp board. Lining the interior surface of the zinc cup is a porous paper separator treated with a mercury salt of limited solubility in water.

Within the cup and in contact with the coated paper is a depolarizing mix, possibly powdered manganese dioxide and carbon. This mix is moistened with an electrolyte—a neutral aqueous solution of zinc and ammonium chlorides. The upright rod is composed of carbon and serves as a positive electrode. The top of this electrode is a conductive metal cap, which serves as a terminal. A conductive top washer made of paperboard forms an expansion space. A heat-fusible seal of wax or pitch rests on the paperboard and seals the top of the cup. A protective cylindrical jacket of cardboard or other nonconductive material encloses the side wall of the zinc electrode.

The mercury salt incorporated in the separator paper has a solubility limited to a maximum of about 0.1 g per 100 g of water at 20°C. Although a salt may have limited solubility in water, it is possible and desirable to have a greater solubility in the electrolyte solution. The quantity of mercury required for satisfactory performance is 0.025 mg/cm^2 of paper.

The chemical action of the separator, which promotes uniform corrosion of the zinc, is believed to proceed as follows: The electrolyte is absorbed from the depolarizing mix into the separator paper. Upon contact, the mercury salt goes into the solution of electrolyte, which makes contact with the zinc; mercury is deposited from the solution onto the zinc and forms an amalgam.

The limited solubility of the mercury salt is important because as elemental mercury is deposited on the zinc, it is removed from solution and more of the mercury salt is dissolved. This action progresses until a protective mercury coating is acquired by the zinc. Contaminating ions that inhibit corrosion must be avoided in the coating operation.

Reflective Insulation

The three forms of heat transfer are radiation, convection, and conduction. In housing, nearly all heat transfer is by radiation or convection. Therefore, any attempts at insulation must be directed toward reducing radiation and convection losses. In the field of radiant heat insulation, aluminum is the most desirable material to use. It is used in the form of foil pasted to paper or as flake coated on paper. Aluminum foil reflects or turns back 95% of the radiant energy that strikes its surface; aluminum-coated paper reflects, 85%. The latter is less expensive and allows moisture vapor to pass through, although if necessary, it can be made into a vapor barrier by laminating with asphalt.

An interesting chemical problem arose when building codes began to specify that aluminum-coated paper be flame retardant. Application of aluminum coating to flame-resistant kraft paper treated with flame-retardant chemicals is now a common practice.

The quality of the aluminum-coated surface and its ability to reflect radiant heat depends on the kind of adhesive and its ratio to aluminum powder, orientation of the aluminum flakes, and the levelness of the surface obtained. In the development of such a product and also its control during production, some method of measuring this reflective property is necessary. Radiation involves three processes: reflectivity, absorption, and emissivity. In other words, the heat that is not reflected by the surface is either absorbed by the material or emitted. Actually, the quality of the heat-reflective surface can be determined by measuring its emissivity and comparing it with that of a black surface, which, under the same conditions, would absorb all the radiation that strikes it and reflect none. Emissivity of a material is a physical property, just as density, color, and hardness are. An accurate method for such a measurement serves as a useful guide during the various steps in the process to insure maximum reflectivity.

Facsimile Copy Papers

Chemical reactivity, heat sensitivity, photosensitivity, and electrophotography have led to the development of a large variety of interesting coated papers.

Diazo Paper

A photosensitive reproduction process can be obtained by coating paper with a thin dry coating of a diazo compound that is a light-sensitive dyestuff. This material is highly sensitive to actinic light, that is, light in the ultraviolet region of the spectrum. Exposure of the coating to actinic light breaks down the diazo compound so that it becomes colorless and incapable of being developed to form a dye. When the part of the diazo compound that has not been exposed to actinic light—for example, the part covered by opaque areas on the original—comes in contact with a developer solution, it combines or couples with the developer to form a dye. Thus, each actinically opaque line of the original is reproduced on the copy. Thus, in those areas that allow actinic light to pass, no color is developed.

These diazo compounds are not affected by exposure to normal room lighting under typical operating conditions. Unlike earlier photocopy systems, this process offers economy, is clean, odorless, and requires no ventilating, inks, stencils, negatives, or special lighting. Developers for this method are available in four colors in addition to standard black. By selection of the colored paper and a color developer, many different combinations are possible.

Xerography

A second process based on electrophotosensitive coating is termed xerography, which is a compound of two Greek words meaning dry and writing. Xerography is a rapid, dry, direct positive electrostatic copying process that uses no water, liquid chemicals, film negative, or dark room. The system is based on the principle of static electricity, by which two objects of unlike electrical charges attract each other.

Single copies can be made by the process or multiple copies can be prepared from an offset paper plate that receives the image from the xerox process. In the original xerox process the heart of the system is the photoconductive selenium surface. This is a transfer intermediate in xerography, just as a negative is in photography. An aluminum plate coated with a selenium surface is positively charged by exposing it to a high voltage corona discharge. The charged surface is able to retain the charge in the absence of illumination and to dissipate selectively the charge when exposed to light or an optical image.

An electrostatic latent image is formed by exposing the sensitized xerographic plate to a light pattern or optical image. The electrostatic latent image is then transformed into a real image by deposition of electroscopic material as a highly colored powder. This powder bears a negative charge, opposite to the charge on the electrostatic latent image and is therefore attracted to the charge areas on the plate. Being opaque the image is not dispelled by exposure to light. The visible image is then transferred to a sheet of paper or to an offset duplicating master.

The sheet of paper is placed over the top of the developed powder image and exposed to the corona discharge. This time the paper takes on a positive charge and attracts the powder. This powdered image is then fused to the final copy.

Electrofax Process

In the electrofax process, developed by RCA, this intermediate step has been eliminated and the paper itself serves in the same way as the selenium-coated aluminum surface in the original process. This method was developed for direct electrophotographic printing on paper and requires a photoconductive coating comprising a special low-cost zinc oxide pigment dispersed in a resin binder and coated on a paper base. A photographic print is produced by placing a uniform electrostatic charge directly on the coated paper, exposing the paper to a light image, causing the charge to drain away in proportion to the intensity of the light incident on the surface, and developing the image by the use of a magnetic brush consisting of a thermoplastic resin powder carried on a brush of iron filings. The coating is insensitive to light unless electrically charged, but attains usable photographic properties when an electrostatic charge is placed on its surface. As in the xerox process, the uniform electrostatic charge on the surface coated with photoconductive resin is accomplished by a corona discharge from an array of fine wires connected to a negative terminal of a 4—7 kV source of direct current. A grounded plate beneath the electrophotographic coating is connected to the positive side of the source.

One interesting consideration in this process is that the paper must be somewhat conductive in order to drain away the electrostatic charge.

Electrically Conductive Papers

The development of these duplicating processes, which require electrically conductive papers, makes the application of conductive organic chemicals to the paper important. To make a paper conductive may be thought of as an extension of the elimination of static electricity on paper by chemical means. Generally speaking, three major mechanisms are involved in antistatic treatment. In one, materials simply increase the hygroscopicity of the fibers. Other chemical compounds, especially ionic surface-active agents, increase

TABLE 4-7

Effect of Chemicals on Conductivity

Additive	Add on, %	12% relative humidity	Resistance, MΩ, at 20% relative humidity	60% relative humidity
Vinyl benzyl quaternary ammonium compound	1.6	144	64	3.2
Quaternary ammonium chloride compound	1.5	18,700	2,130	10.3
Nonpolymeric quaternary ammonium salt of moderate molecular weight	2.1	37,500	23,100	5.7
Potassium chloride	1.8	—	27,400	46.5
Control	—	37,500	37,500	17,600

Note: Tests were carried out on sulfite bond paper with rosin size. Add on is the percentage of nonvolatile added, based on the weight of the dry paper.

the surface conductivity of the fibers and speed the dissipation of the charge. Still other materials tend to induce an opposite charge to the one that normally builds up on the fiber itself. This static charge can be positive or negative, depending upon the orientation of the molecules in the outer layer of the surface-active agent.

Particularly in demand are papers of a greater electrical conductivity than ordinary paper and that retain their conductivity over a wide range of humidity. Ordinary untreated paper has a low electrical conductivity especially when it is dry. At a relative humidity of 12% the resistance of untreated paper is more than 37,500 MΩ. Inorganic salts, such as potassium chloride or other conductive materials, can be added to the paper to increase the conductivity. Surface treatment or impregnation with chemicals is accomplished by conventional coating techniques.

Inorganic salts, although widely used, do not provide satisfactorily high conductivity at low humidities even with the aid of humectants. Carbon black produces suitable electrical properties at the expense of disagreeable appearance and excessive opacity for most applications.

Organic chemicals, such as polyvinylbenzl–trimethyl ammonium chloride, lower electrical resistance to 10^6—10^9 Ω when applied to paper.

Table 4-7 illustrates the effect of such materials on the conductivity over a broad range of relative humidity.

Pressure-Sensitive Manifold Copy Paper

Pressure-sensitive manifold copy paper is a transfer sheet useful in duplicating and manifolding operations such as business forms and sale contracts. Notable among the papers supplied with a transfer coating is one developed by National Cash Register Company known as NCR paper. The system depends on two coated surfaces, one that acts as a donor or transfer surface and another that acts as a receiver or acceptor of printing fluid. This transfer coating consists of a hydrophilic colloid throughout which a profuse number of microscopic droplets of an oil-marking fluid are dispersed. Each droplet of oil is encased in its own capsule of the colloid material and is actually produced, not from an emulsion, but from a process of coacervation. This transfer film is rupturable by the pressure of marking, in order to release the marking fluid from the hydrophilic colloid. In a transfer sheet that carries a coating of these capsules, and as pressure is applied by a writing or printing instrument, the wall of the hydrophilic capsule is ruptured at the points of pressure and the fluid is transferred to the undersheet.

The colorless dyes used in the NCR system are crystal-violet lactone and benzoyl leucomethylene blue. These oil-soluble dyes are dissolved in a combination solvent of Aroclor—a chlorinated diphenyl and magnaflux oil. The resulting colorless oil solution is stable in light and air. When this solution comes into contact with a naturally occurring clay—e.g., attapulgite—the color block on the dyes is chemically removed. The clay acts as a strong acid and chemically ruptures a bond on each of the dyes. The formation of crystal-violet lactone color is instantaneous. However, the resultant print is somewhat fugitive to light. The benzoyl leucomethylene blue forms its color much more slowly, but its print is resistant to fading.

Encapsulation

To use this colorless ink in a paper coating operation requires protection of the oil. Such protection was provided by the microencapsulation technique developed at NCR. By this technique, a colloid film composed of gelatin and gum arabic is automatically formed around each and every drop of the colorless dye solution. The resultant capsules are approximately 1—2 μ in diameter, with aggregates of 8—15 μ. The system might be likened to a bunch of grapes, which is actually the way its appears when viewed under a microscope. The seeds might be considered as the dyes, the jell material as the solvent, and the skin on each grape as the encapsulating protective wall. Each grape then would be approximately 1—2 μ in diameter and bunches or aggregates would form in sizes of 8—15 μ.

The hardened capsule around wall each oil drop is obtained by tanning with formaldehyde. The wall completely protects the oil, enables it to be treated as a solid material, and allows it to be coated on conventional equipment. The capsule wall is then ruptured by pressure of writing, the oil solution is transferred to the clay coating and a print immediately develops.

Self-Contained Record Paper

In new self-contained record paper, both the capsules and the clay are coated on the same side of the sheet. Obviously the problems in preventing premature development of color are manifold in such a system. Yet it may well gain wide acceptance in areas where inking systems might be eliminated. Examples of this application are chart recorder papers or data processing paper for high-speed computers.

One of the present problems with high-speed computers is the inability to produce a satisfactory inking system that can keep up with the machinery and not cause ribbons to be damaged under high-speed operation. Because it produces color when pressure is applied, the transfer film must have a minimum threshold pressure to resist destruction by rough handling so as not to smudge the paper prematurely. Consequently, additives must be put into the system to build threshold pressures, control the area of contact, and provide the necessary antifriction qualities.

Besides being a replacement for carbon, the capsular system can be employed in a number of interesting applications in the paper and other fields. Encapsulating techniques might be extended to adhesives, volatile aromatic fluids, cleansing solutions, and ink transfer products.

Electrosensitive Recording Papers

Trends in modern technology have been placing increasing demands on fast, convenient, and inexpensive methods for printing line copy or continuous tone images and also for recording coded information. High speed printing of the output of electronic computers is limited by relatively slow recording methods when compared with the computation rate of the machine. Recording speech and data from high-speed aircraft and satellites in orbit is assuming considerable importance.

Paper impregnated with electrically conductive solutions that decompose by electrical impulse are able to produce a record. An electric arc can also be caused to burn holes in a paper recording blank and thus create outlines of the subject matter to be recorded. A widely-used type consists of carbon-impregnated conductive paper with a metallic pigmented coating on one side.

Electrosensitive recording papers are conventionally classified as "wet" and "dry." Dry recording papers operate on the breakdown principle, in which sufficient voltage must be supplied to break down the insulating layer of coating before the paper is marked. The localized action of an electrical stylus produces a chemical decomposition of the surface coating, which in turn produces a conspicuous show through of a black base material. The breakdown voltage in conventional dry electrosensitive recording papers usually exceeds 75 V. Wet electrosensitive papers are believed to operate at lower voltages.

The base paper usually contains carbon, graphite, or other conducting material, such as metallic powders, to produce a uniformly high conductive base paper. The coating usually contains a hygroscopic adhesive in which a conducting salt and a metallic powder are dispersed.

Outlook For Coating Papers

The reasons for coating paper have extended far beyond the simple requirement to improve the printing surface and add a decorative effect. Functional demands, in addition to printing, embrace the fields of packaging, building construction, plastics, office copying, duplicating, multiple sales forms, and information recording. The fast-moving market requirements for new kinds and large quantities of coated papers have stimulated major technological advances in methods for coating preparation, material handling, coating application, drying, web handling, and finishing. Each unit process is advancing to a sophisticated level of production, utilizing instruments and automation wherever possible. The coating industry has grown into an important segment of the paper industry, adding utility and value and satisfying a market demand. The products described cover a small segment of the total industry, yet it serves well to record some of the past activity, and point toward possible greater interest and accomplishments for the future.

REFERENCES

1. O. Kress and C. Johnson, "Method For Coating Paper", US Patent 2,339,707 (1944).
2. C. Prutton and S. H. Maron, *Fundamentals of Physical Chemistry*, Macmillan, New York (1947).
3. N. C. Cahoon, US Patent 2,534,336 (1950).
4. J. J. Coleman, US Patent 2,598,226 (1952).
5. J. B. Conant and A. H. Blatt, *The Chemistry of Organic Compounds*, Macmillan, New York (1952).
6. H. R. Kruyt, *Colloid Science*, Vol. I, Elsevier, New York (1952).
7. American Cyanamid Company, *Unitane Pigments Handbook* (1956).
8. P. Becher, *Emulsions* Reinhold, New York (1957).
9. J. R. Belche and G. C. Ellis, "Viscosity Control of Paper Coating Adhesives with Urea," *Tappi*, **40** (?) (1957).
10. G. A. Hemstock and J. W. Swanson, "A Study of the Penetration of Coating Color Components by Means of the Roll-Inclined Plane Technique", *Tappi*, **40** (10) (1957).
11. C. A. Richardson, *Tappi*, **40** (8) (1957).
12. J. J. Bickerman, *Surface Chemistry*, Academic Press, New York (1958).
13. R. M. K. Cobb, "Coating Adhesive Demand—What Pigment Function Governs It?" *Tappi*, **41** (10) (1958).

14. J. C. Stinchfield, R. A. Clift, and J. J. Thomas, "The Water Retention Test in Evaluating Coating Color," *Tappi*, **41** (1958).
15. *Tappi* Monograph Series No. 20, "Paper Coating Pigments" (1958).
16. W. R. Willets and F. R. Marchetti, "Effect of Pigment Volume Concentration on Coating Properties," *Tappi*, **41** (12) (1958).
17. R. I. Hagerman, R. G. Jahns, and W. H. Somers, "A Study of the Water Retention of Latex-Bound Pigmented Coated Colors," *Tappi*, **42** (9) (1959).
18. J. V. Robinson, "The Dispersion of Pigments for Coating Color: Concept and Theory," *Tappi*, **42** (6) (1959).
19. M. H. Vander Velde, *Tappi*, **42** (11) (1959).
20. R. L. Whistler and J. N. BeMiller, *Industrial Gums*, Academic Press, New York (1959).
21. A. C. Eames, "The Transverse Tensile Strength of Clay-Starch Coating," *Tappi*, **43** (1) (1960).
22. T. A. Gardner, "A Theory of Drying With Air," *Tappi*, **43** (9) (1960).
23. N. Millman and J. B. Whitley, "The Effect of Mixing Intensity on the Dispersion of Coating Pigments in Water," *Tappi*, **43** (12) (1960).
24. R. E. Weber, "The Effect of Pigment Binders on Paper Gloss, Printed Gloss, and Oil Absorption," *Tappi*, **43** (10) (1960).
25. J. P. Casey, *Pulp & Paper*, 2nd ed., Vol. III, Interscience, New York (1961).
26. R. A. Daane and S. T. Han, "An Analysis of Air-Impingement Drying," *Tappi*, **44** (1) (1961).
27. L. E. Georgevits and W. L. Marino, *Tappi*, **44** (10) (1961).
28. T. L. Reilling, "Pigment Paper Coating Composition," US Patent 3,002,844 (1961).
29. *Tappi* Monograph Series No. 22, "Synthetic and Protein Adhesives for Paper Coating" (1961).
30. N. J. Bechman, "The Wet Resistance of Coated Offset Papers," *Tappi*, **45** (11) (1962).
31. A. Harsveldt, "Starch Requirements for Paper Coating," *Tappi*, **45** (2) (1962).
32. E. J. Heiser, R. W. Morgan, and A. S. Reder, "Cationic Latex with Controllable Penetration and Setoff Properties," *Tappi*, **45** (7) (1962).
33. G. Kosta and A. W. Willenbrock, *Tappi*, **45** (9) (1962).
34. L. I. Osipow, ACS Monograph 153, *Surface Chemistry: Theory and Industrial Applications*, Reinhold, New York (1962).
35. R. R. Rounsley, *Tappi*, **45** (2), 154A-156A (1962).
36. H. C. Schwalbe and C. E. Libbey, *Pulp & Paper Science and Technology*, Vol. II, McGraw Hill, New York (1962).
37. "Microencapsulation," *Management Reports*, Boston (1963).
38. *Tappi* Monograph Series No. 25, "Paper Coating Additives" (1963).
39. G. Feig, "Electrostatic Printing," *Pulp & Paper*, **38** (29) (1964).
40. J. J. Clancy *et al*, "Starch Stabilized Casein Coating Emulsions," US Patent 3,157,533 (1964).
41. E. J. Heiser and D. W. Cullen, *Paper Trade J.* **140** (20) (1964).
42. *Tappi* Monograph Series No. 28, "Pigmented Coating Processes For Paper and Board" (1964).
43. Baxter L. Willey, *Pulp & Paper*, **38** (41) (1964).
44. C. J. Heiser and D. W. Cullen, "Effects of Drying Rates on Adhesive Redistribution of Coated Papers Properties," *Tappi*, **48** (8) (1965).
45. H. H. Morris, P. Senett, and R. J. Drexel, "Delaminated Clays—Physical Properties and Paper Coating Properties," *Tappi* Coating Conference (1965).
46. *The Dictionary of Paper*, 3rd ed., American Paper and Pulp Association, New York (1965).

47. *The Paper Year Book*, Ojibway Press, Duluth, Minn. (1965).
48. K. H. Rosenstock, "Mill Practice in The Use of Coating Pigments in Europe," *Tappi* Coating Conference (1965).

chapter 5

Solvent Coatings and Solvent-Recovery Systems

M. C. BEREN

Solvent coating solutions contain nonvolatile, organic film-forming substances dissolved or dispersed in one or more volatile organic liquids. The nonvolatile organic substance may be a single component or a mixture of several substances capable of forming a continous film on evaporation of the volatile liquid. The volatile liquid is called the solvent. These systems are known as lacquers in trade terminology.

HISTORY

The development of solvent coatings for paper has in general followed the development of film-forming materials and solvents very closely. The first solvent coatings for paper were probably the so-called spirit varnishes, which were usually solutions of natural copal resins in alcohol. Their widest application was to sheets, such as labels and advertising literature. They produced good gloss in clear coatings, but had several inherent drawbacks. The coatings were brittle and had a tendency to powder when folded; they softened and blocked in warm weather and, in general, had poor mechanical properties and poor chemical resistance.

The solvent-coating industry really began to develop after World War I, when soluble nitrocellulose and solvents became available in quantity. The early 1920's brought a substantial quantity of paper coated with nitrocellulose and embossed in grains and fancy designs to simulate leather. These coatings were usually made on kraft paper or on a heavier, saturated paper. They found use in many industries: for sock lining and heel pads in shoes, picture-frame covers, box covers, bookbindings, and all sorts of novelties. Almost all these coatings were made with nitrocellulose or nitrocellulose scrap from celluloid, film, or smokeless powder left from the war.

The development of fancy coatings on paper, with solvents, began in the early 1930's. Among these were crystal, high-gloss, and metallic papers. Nitrocellulose still played the leading role in these coatings and little or nothing was yet done with the other cellulose derivatives in this field. For the most part, these fancy finishes were used primarily for their pleasing decorative effects, although they also served a utilitarian function by virtue of their water resistance and durability.

About the same time, much interest was shown in clear solvent coatings for applications other than decorative, in which resistance to water, water vapor, grease, foods, and drugs was the chief aim. Much work was done with synthetic resins—both phenolics and alkyds—in combination with nitrocellulose, to obtain these properties. In particular, large quantities of glassine were coated with some of these combinations to produce functional finishes.

The development of solvent coatings made with other cellulose derivatives began about the middle of the 1930's. Cellulose acetate, mixed cellulose esters, and ethyl cellulose were investigated and used in increasing quantities. The production and use of synthetic resins grew rapidly. About this time, the use of vinyls for paper coatings assumed some importance. The growing need for suitable plasticizers and solvents for these newer materials was met by the increasing developments coming out of research laboratories throughout the country.

By World War II, the solvent-coating formulator had a wide range of coating materials from which to choose, such as cellulose derivatives and synthetic resins of many types, including vinyls. Although many civilian uses for decorative and protective papers were curtailed by the war emergency, the requirements of the military services for coated papers brought about the expansion of existing paper-coating facilities. The global war, being fought in all sorts of weather conditions, demanded coatings that could withstand extremes of heat and humidity.

Rigid specifications were set up for many kinds of military packaging, in particular, for wrapping ordnance parts. These specifications included resistance to water, water-vapor, and grease, flexibility at low temperatures, and resistance to blocking at high temperatures. In many cases, these specifications were hurriedly drawn and required coating properties impossible to obtain. This condition, typical of the waste of war, could hardly be avoided. The specifications in many cases were far beyond the limitations of nitrocellulose, and increasing attention was given to the various types of vinyls and other synthetic resins, cellulose acetate, and ethyl cellulose. Extensive development work was also done on coatings for paper that could resist military gases. Fortunately, the actual use of this type of paper did not become necessary during the war.

The manufacturers of solvent-coated papers had hopes that many of these special papers, which had been used for ordnance wraps, would find wide application in civilian industries after the war for wrapping many kinds of machine parts. These expectations, however, did not materialize to any great extent.

Since the war, about the same types of decorative and protective paper are available as before the war. Nitrocellulose is used as extensively as ever for paper coatings, but the use in this field of newer types of coating materials such as vinyls and other synthetics has greatly expanded.

COMPARISON WITH AQUEOUS COATINGS

The general characteristics of solvent coatings can be advantageously described by comparing them with those of aqueous coatings, from the following points of view.

Mechanism of Drying

In both types of coating, the entire mechanism of drying is usually by the evaporation of the solvent. In aqueous coatings, it is the water that evaporates and deposits the nonvolatile ingredients, such as casein, in a continuous film. In solvent coatings, it is the organic solvents that evaporate. In both cases, the coating may contain substances that are still tacky after the evaporation of the solvent and require further treatment to become converted to a dry state. These compounds may be thermosetting synthetic resins or compounds that undergo further polymerization, condensation, or oxidation, at elevated temperatures. By far the largest commercial applications of solvent coatings for paper involve only solvent evaporation as the mechanism of drying.

Speed of Drying

One of the chief advantages of solvent coatings is their speed of drying. With the solvent combinations commonly used in paper coatings, the drying is much more rapid than that of aqueous coatings. From the practical point of view, this means that the drying equipment for solvent coatings can be smaller and, consequently, less expensive. It also means that under the same drying conditions, solvent coatings can be run at higher speeds.

Methods of Application

Solvent coatings are generally applied either by one of several modifications of doctor knife or by means of some type of roll coater. Other kinds of coaters can also be used, such as spray, air brush, or print types, but most solvent coatings used today are applied by the first two means. Aqueous coatings are generally applied by a brush coater or by an air brush. Knife coaters and roll coaters are used for some applications.

The advantages and disadvantages of various types of coating equipment are discussed in another section of this chapter.

Properties of the Resulting Films

In the final analysis, the chief interest in comparing solvent coatings with aqueous oatings lies in the comparison of the properties of the dry films. Here lie the greatest differences, to the advantage of the solvent types. In general, solvent coatings are much more water and vapor-resistant. They have better mechanical properties: tensile strength, flexibility, and resistance to handling. They are more resistant to many types of food product and drugs and to chemical and bacterial action. In the decorative field, solvent coatings can be formulated to produce a much greater variety of finishes, and they can be made with higher gloss and greater depth of color free from mottling. These statements may have to be modified in the future, because special types of aqueous vinyl dispersions are becoming available. These show promise of producing finished films that may equal some of the solvent types in many respects.

Hazards of Application

Fire, explosion, and toxicity hazards are important considerations in the application of solvent coatings on paper. All the commonly used organic solvents are highly inflammable and form explosive mixtures with air within certain ranges of concentration. Fire and explosion must be guarded against in all phases of manufacture, from the handling of the raw materials to the final coating and drying operations. In addition, special precautions must be taken where nitrocellulose is handled. Nitrocellulose is normally supplied wetted with 35% denatured ethyl alcohol. In this state, it is safe to handle with reasonable precautions. However, if allowed to dry, it becomes a highly inflammable material—a condition that must be avoided.

Although, from the toxicity viewpoint, most of the commonly used solvents are safe when reasonsble ventilation is provided, every solvent has limits of concentration beyond which it is definitely narcotic or toxic to the human body. This toxicity may be either acute or cumulative, depending on the solvent.

These problems will be discussed in some detail later in the chapter.

Economics

Solvent coatings are usually more expensive than aqueous coatings. This is obvious from the fact that the cost of water is negligble, whereas the cost of solvents is relatively high. In almost all cases, the cost of the nonvolatile components of the solvent coating is higher than that of aqueous coatings. The combination of the high cost of the solvents and the generally higher cost of the nonvolatiles results in a higher cost per unit volume of dry coating deposited. Rarely, the cost of the nonvolatile ingredients in an aqueous coating causes a higher total cost per unit volume of solids deposited than that of some solvent coatings.

A word should be said here about the effect from the economic point of view of solvent recovery on the comparison of solvent coatings with aqueous coatings. Where the load is sufficiently high to warrant the investment, a properly designed and efficiently operated recovery plant can show great savings. However, even with the savings that can result from an efficient recovery plant, solvent coatings still are more costly than aqueous coatings.

PURPOSE OF COATINGS

Papers are coated to add certain desirable properties that uncoated papers lack or do not possess in sufficient degree. In general, these are:

> Pleasing appearance and feel
> Resistance to water, grease, chemicals, drugs, and foods
> Resistance to the transmission of water vapor and gases
> Physical resistance to scuffing, abrasion, and folding
> Special properties, such as heat sealing

In accomplishing one or more of these major aims, the coating should have some or all the following important properties:

> Freedom from odor, taste, and toxicity
> Resistance to discoloration by sunlight
> Nonblocking under reasonable pressure and temperature

Reasonable cost

Good adhesion to the base

Low inflammability

Stability to aging

No single coating can be expected to accomplish all these aims or have all these properties to the maximum degree. However, some of the properties are indispensable for any conceivable use. For example, unless a coating is nonblocking under the conditions of manufacture and use, it is worthless for any application. The same holds, almost to the same degree, for adhesion. If the coating peels easily, its value is limited. In a purely decorative paper, the principal aim is pleasing appearance. If, in addition to its decorative appeal, it has other desirable properties, so much the better; but it may not need to have high resistance to water-vapor transmission, to grease, or to abrasion. Some of these properties are occasionally conflicting, and a formulation for a specific use should always represent an optimum compromise. This problem will be discussed further under Formulation.

TYPES OF SOLVENT-COATED PAPER

For purposes of classification, solvent-coated papers can be placed into two general groups: decorative; protective and functional.

No sharp line of demarcation exists between the two. Papers of both types can serve a dual purpose, and can be both decorative and protective at the same time. However, for simplicity, a coated paper is classified here as decorative if its main appeal is in its appearance, and as functional if its main purpose is either protection of some kind or use other than decorative.

Decorative Papers

Decorative papers depend largely on surface finish.

High-Gloss or Patent-Leather Paper

This is a coated paper with a glossy surface that can be made in many colors. Although its special appeal is in its high finish, it also serves as a waterproof and soilproof surface for the protection of packaged materials. It has its counterpart, (in the aqueous coated paper field,) in flint papers, and in coated papers that are dried against a highly polished drum.

Matte-Finish Paper

This type of coated paper has a dull or flat finish, and its chief appeal is in its velvety sheen. Like the high-gloss paper, it is primarily decorative, but it serves a protective function as well.

Metallics

These are the well-known coatings that contain metallic powders, such as bronze, copper, or aluminum, to produce gold and silver effects. The coatings can be made in a variety of metallic shades, depending on the composition of the bronze powder used. By introducing spirit-soluble dyes into the coating, many attractive colored metallic shades can be made.

Crystal Paper

This paper has the appearance of a crystal pattern and is made by adding certain organic crystals, such as phthalic anhydride, to the coating solution. The pattern depends on the type of crystal used and very beautiful effects can be obtained. Some of the most striking patterns, unfortunately, cannot be used for packaging and similar purposes, because of the odor of the particular crystal used.

Iridescent Coating

This is a relatively new decorative for paper. It has the multicolored appearance of the spectrum colors seen in a film of oil on water or in a soap bubble. Embossing increases the iridescent effect through mutual reflection of the embossed facets.

Flock Finish

This finish has a fibrous, velvety effect, obtained by depositing short colored fibers on a tacky coating. Several kinds of fibers can be used, and the effect depends somewhat on the type of the fiber.

Mother-of-Pearl Finish

This finish is usually made by adding pearl essence, obtained from fish scales, to the coating solution.

Grain and Marble Finish

This is a simulated wood-grain or marble effect and can be made by coating with an imperfect mixture of several colors under controlled conditions.

Embossed Papers

All the types of paper mentioned can be embossed in various grains and patterns. Hundreds of designs are available and embossed solvent-coated papers have been used for years for many kinds of applications.

Two-Tone or Spanish Finish

This type of finish can be made by embossing a colored coating, and then "souping," i.e., covering the surface with a thin solution of a contrasting color. If the souping operation is done by a doctor blade, the contrasting color remains in the crevices. If it is done with a roll, the contrasting color can be made to hit only the high spots. The second procedure is commonly called "topping."

Protective and Functional Papers

Coated papers belong to the class of protective and functional papers. Their main functions are to provide:

Resistance to water and dirt
Resistance to water-vapor transmission
Resistance to grease
Resistance to specific materials such as foods, drugs, and chemicals
Heat sealing

A functional coating may have more than one of these properties to some degree. There are no universal methods of testing, or specifications, to determine whether a coated paper is waterproof or vaporproof; these are relative terms. Many tests and specifications for these functions have been suggested and used. The problem of devising a laboratory test that would predict accurately how a coated paper will behave in practice is not a simple one. Many tests havs been set up, and their conclusions have been at great variance with actual results, because they neglected important factors. Other specifications and tests have been extreme and have required test values beyond those reasonably needed. No set of laboratory tests can ever duplicate actual use. They give, however, a comparative picture of the properties being studied, and act as a guide to further improvement. Wherever possible, the standard tests set up by TAPPI and the ASTM should be used.

Artificial Leather

Artificial leather paper cannot be classified either as decorative or functional in the strictest sense of these terms. The chief aim of this product is to simulate the appearance, feel, and utility of real leather. The best types are made by coating a paper that has been saturated with natural or synthetic rubber latex. The coating is usually made from nitrocellulose or vinyls, and is embossed with a grain to give the appearance of leather. The coating must be water-resistant, abrasion-resistant, and flexible, and must adhere well to the base.

USES OF COATED PAPERS

Uses of solvent-coated papers have already been outlined in this chapter. However, at the risk of some repetition, a list of the many uses and fields of application is given as follows:

Packaging

Box covers
Containers and bags
Labels

Display

Window, counter, and booth displays

Printing Trade

Catalogue covers
Menu covers
Advertising literature
Greeting cards

Household

Shelf papers and linings for drawers
Waterproof table covers and doilies
Washable window shades
Washable wall paper
Lamp shades

Imitation Leather on Paper Base

Socklining, heel pads, and underlay for shoes
Brief cases and bags
Photograph albums and scrap-book covers
Novelty items, such as comb cases and wallets

Industrial

Gaskets for special applications
Electrical insulation
Adhesive papers

Specialties

Decalcomania
Special photographic papers
Imitation flowers

This list is by no means complete, but it serves to illustrate the wide variety of uses for coated papers.

INGREDIENTS OF SOLVENT-COATING SOLUTIONS

Organic film-forming materials are primary or secondary. The nonvolatile portion of a solvent coating solution may occasionally be a single substance, but it is usually a mixture of several substances. Its ingredients may be divided into the following general classes:

Primary film-forming materials
Secondary film-forming materials
Plasticizers
Pigments, dyes, fillers, and bronze powders

Primary Film Formers

The distinction made here between primary and secondary film-forming materials is somewhat arbitrary. In general, however, the primary film-forming components make up the principal binding portion of the film, and as a rule, they have better mechanical properties than the secondary film formers. Chemically, they are polymers of high molecular weight. They form the nucleus around which the film is built, and they probably contribute most to its properties. Secondary film formers are resins (other than those classified here as primary film formers) used to modify the properties of the primary film formers. Coatings can be made from resins alone or from resin and plasticizer combinations, but they have many shortcomings for most applications and are not considered here.

An ideal primary film former should have the following properties:

Physical

High tensile strength and elasticity
Good resistance to abrasion
Good solubility in readily available organic solvents
The ability to be plasticized easily
Low moisture absorption
Insolubility in water
Freedom from odor
Compatibility with a wide range of other film-formers
Low solvent retention
Freedom from toxicity and adverse physiological reactions
Resistance to burning
Reasonable cost

Chemical

Stability against decomposition under conditions of use

Resistance to oxidation

Resistance to discoloration and decomposition by heat and sunlight

Nonreactivity (except thermosetting) with the other components of the coating

Resistance to acids and alkalies, and to other agents if this is required in the specifications

None of the following materials has all these properties, but the more closely a material approaches this ideal, the more suitable it becomes for paper-coating purposes.

The following primary film formers, which are commercially available, will be considered here:

Nitrocellulose

Cellulose Acetate

Cellulose Acetate Butyrate

Cellulose Acetate Propionate

Ethyl Cellulose

Benzyl Cellulose

Polyvinyl Acetate

Polyvinyl Chloride

Copolymer of Vinyl Chloride and Vinyl Acetate

Copolymer of Vinyl Chloride and Vinyl Acetate

Polyvinyl Butyral

Copolymer of Vinyl Chloride and Vinylidene Chloride

Copolymer of Vinylidene Chloride and Acrylonitrile

Polyethylene

Chlorinated Rubber

Cyclized Rubber

Zein

Most of these compounds are described briefly, with regard to manufacture, chemical structure, solubility, and general properties, so that so that their applications to paper coatings can be evaluated more intelligently. Where compatibility data are given, it must be understood that compatibilities often depend on the presence of other substances, such as plasticizers and resins and that, under some conditions of formulation, discrepancies are possible.

Nitrocellulose

Chemical Constitution. Nitrocellulose is the nitric acid ester of cellulose. It is made from purified cotton linters or wood pulp by controlled
nitration with a large excess of a mixture of nitric and sulfuric acids.
Fully nitrated cellulose, or cellulose trinitrate, has a theoretical nitrogen
content of 14.14%. This is not attainable in practice, and the upper
limit is about 13.8%. The nitrocellulose used for coating lies in the
range 10.7—12.2% nitrogen, with most of it in the range 11.8—12.2%.
This means that some of the hydroxyl groups in the anhydroglucose
units are unreacted. A fully nitrated cellulose may be represented by
the following structural formula:

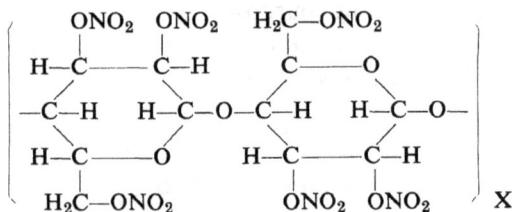

$$
\left[
\begin{array}{ccc}
\overset{\displaystyle ONO_2}{\underset{|}{}} \;\; \overset{\displaystyle ONO_2}{\underset{|}{}} & \overset{\displaystyle H_2C-ONO_2}{\underset{|}{}} & \\
H-C\!\!-\!\!-\!\!-\!\!C-H & C\!\!-\!\!-\!\!-\!\!O & \\
-C-H \;\; H-C-O-C-H & H-C-O- \\
H-C\!\!-\!\!-\!\!-\!\!O & H-C\!\!-\!\!-\!\!-\!\!C-H \\
\underset{|}{} & \underset{|}{} \;\; \underset{|}{} \\
H_2C-ONO_2 & ONO_2 \;\; ONO_2
\end{array}
\right]_X
$$

Effect of the Nitrogen Content on the Properties of Nitrocellulose. Both
the nitrogen content and viscosity can be varied, with resulting effects
on the solubility and other properties. As the nitrogen content is reduced to the lower limit, the alcohol tolerance increases. The tolerance
for hydrocarbon diluents also varies with the nitrogen content. Depending on the active solvent and the type of hydrocarbon used, the
tolerance may increase progressively with increase in nitrogen content,
or it may reach a maximum somewhere between the lower and upper
limits of nitrogen content. The tensile strength and elongation rise
with an increase in the viscosity of the nitrocellulose.

The commercially available types of nitrocellulose are listed in Table
5-1.[1]

The lower-viscosity types of SS nitrocellulose are almost completely
soluble in ethyl alcohol alone and are completely soluble in mixtures of
ethyl alcohol and aromatic hydrocarbons.

The RS type of nitrocellulose is the one most commonly used; it
has the greatest range of solubility.

The AS type of nitrocellulose is soluble in the same active solvents,
but alcohol is a more satisfactory diluent for it than are the hydrocarbons.

The solvent mixture used in the viscosity measurements of Table 5-1

TABLE 5-1

Nitrogen and Viscosity Specifications of
Soluble Nitrocellulose Types

Designation	% Nitrogen	cP, of a solution of 12.2% concentration	Viscosity		
			sec, of a solution of		
			12.2%	20%	25%
SS ¼ sec					4-5
SS ½ sec	10.7-11.2			3-4	
SS 5-6 sec			5-6.5		
AS ½ sec	11.2-11.7			3-4	
RS 10-15 cP		10-15			
RS 18-23 cP		18-23			
RS 30-35 cP		30-35			
RS ¼ sec					4-5
RS ½ sec				3-4	
RS 5-6 sec	11.8-12.2		5-6.5		
RS 15-20 sec			15-20		
RS 30-40 sec			30-40		
RS 60-80 sec			60-80		
RS 125-175 sec			125-175		
RS 250-400 sec			250-400		
RS 600-1000 sec			600-1000		

TABLE 5-2

General Properties of Nitrocellulose [1]

Appearance	White fibers
Color	Water white
Odor and Taste	None
Specific Gravity	1.65
Refractive Index	1.514
Light Transmission	Transparent down to 3130 Å (absorbs sunlight)
Effect of Sunlight	Considerable discoloration and embrittlement in clear film
Water Absorption (80% RH for 24 hr)	1.0%
Chemical Resistance	Fair resistance to weak acids; poor resistance to strong acids, and to weak and strong alkalis; good resistance to hydrocarbons and oils
Effect of Heat	Decomposes slowly at 135°C
Flammability	Extremely inflammable

is composed of 25% specially denatured No. 1 ethyl alcohol, 20% ethyl acetate, and 55% toluol, all by weight.

Solubility. Nitrocellulose is soluble in a wider range of solvents than any of the other cellulose derivatives except ethyl cellulose. It is soluble in many esters, ketones, and ether alcohols. Alcohols alone do not dissolve it, but they take on solvent properties when used together with active solvents. Aromatic and aliphatic hydrocarbons can serve large proportions as diluents. Some of the most common solvents and diluents are shown in Table 5-3.

TABLE 5-3

Common Cellulose Nitrate Solvents and Diluents

Solvents	*Latent solvents*
Methyl Acetate	Ethyl Alcohol
Ethyl Acetate	Isopropyl Alcohol
Isopropyl Acetate	Butyl Alcohol
Butyl Acetate	Amyl Alcohol
sec-Butyl Acetate	
Amyl Acetate	
Ethyl Lactate	*Diluents*
Acetone	Benzol
Methyl Ethyl Ketone	Toluol
Cellosolve	Xylol
Methyl Cellosolve	Petroleum Hydrocarbons
Cellosolve Acetate	
Methyl Alcohol	

With a given type and concentration of nitrocellulose, the viscosity of the resulting solution varies greatly with the solvent or solvent combination used. As an illustration, Table 5-4 shows the viscosity of a 10% concentration of RS ½ sec nitrocellulose in various solvents.

TABLE 5-4

Effect of Solvents on Viscosity [1]

Solvent (commercial grade)	*Viscosity, cP at 25°C*
Methyl Acetate	34
Ethyl Acetate (99%)	38
Ethyl Acetate (85-88%)	33
n-Butyl Acetate	62
n-Amyl Acetate	83
Hexyl Acetate	122
Octyl Acetate	281
Ethyl Lactate	287
Butyl Lactate	328

Compatibility. Nitrocellulose is compatible with nearly all types of chemical plasticizers and with many oils. Scores of plasticizers can be used, but only a few of the most common types are listed in Table 5-5.

TABLE 5-5

Common Plasticizers for Cellulose Nitrate

Castor Oil	Tricresyl Phosphate
Blown and Special Processed	Tributoxy Ethyl Phosphate
Castor Oils	Tributyl Phosphate
Dibutyl Phthalate	Butyl Acetyl Ricinoleate
Diamyl Phthalate	Methoxy Ethyl Acetyl Ricinoleate
Dioctyl Phthalate	Paraplex RG 2
Dimethoxy Ethyl Phthalate	Paraplex G 20
Dibutoxy Methyl Phthalate	Paraplex G 25
Diallyl Phthalate	

Nitrocellulose is compatible with nearly all types of resin both natural and synthetic.

Nitrocellulose is compatible with ethyl cellulose, cellulose acetate, the mixed cellulose esters, vinyl acetate, and polyvinyl butyral. It is incompatible with polyvinyl chloride, the copolymer of vinyl chloride and vinyl acetate, the copolymer of vinylidene chloride and acrylonitrile, chlorinated rubber, and cyclized rubber. It has only very limited compatibility with the copolymer of vinylidene chloride and vinyl chloride.

Cellulose Acetate

Chemical Constitution. Cellulose acetate is the acetic acid ester of cellulose, and is made by reacting purified cellulose with acetic anhydride in the presence of acetic acid as a solvent.

If all three hydroxyl groups in each anhydroglucose unit are reacted, the triacetate is formed, with a content of 62.5% combined acetic acid. For most commercial applications, however, cellulose acetates of lower acetic acid content are more suitable. The triacetate is therefore subjected to hydrolysis to obtain compounds of acetic acid contents ranging from 51.0 to 58.6%. This corresponds to a 37—42% acetyl content.

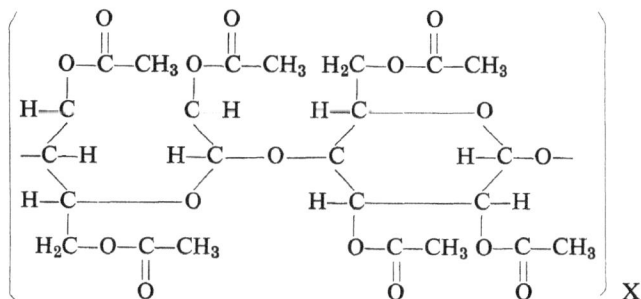

The triacetate may be represented by the following structural formula:

Acetyl Content and Viscosity. Because the acetyl content and the molecular weight of the cellulose acetate polymer can be varied over a wide range, several types are possible. The acetyl content and viscosity determine the properties of the ester to a great extent. The acetyl content has a direct bearing on the solubility; in general, the lower the acetyl value the wider the range of solubility. In particular, lowering the acetyl content increases the tolerance for alcohol. It also tends to make the acetate more thermoplastic. Viscosity has an effect on solubility, and for the same acetyl content, the lower the viscosity of the ester the wider is the range of solubility. Like other polymers, higher-viscosity acetates produce films of greater strength and flexibility than lowerviscosity acetates.

Available Cellulose Acetate Types. The types of cellulose acetate that are most likely to be suitable for paper coatings are listed by

TABLE 5-6*

Viscosity and Acetic Acid Specifications of Commercial Cellulose Acetates According to One Manufacturer

Type	Viscosity, sec	% Combined acetic acid
†A-48	2-5	55.3-56.3
A-1	5-8	55.3-56.3
A-9	15-30	55.8-56.8
A-13	50-80	53.5-54.8
†E-4	8-15	53.5-54.8
E-10	30-50	53.5-54.8
E-13	50-80	53.5-54.8

† E type designates granules, A type designates flake. The E type is more readily dispersed in solvents.

TABLE 5-7*

Viscosity and Acetic Acid Specifications of Commercial Cellulose Acetates According to Another Manufacturer

Type	Viscosity, sec	% Combined acetic acid
LL-1	2-4	55.0-56.0
LM-1	25-50	54.0-55.0
LH-1	60-80	54.0-55.0
FM-3	20-25	56.1-56.6
FM-6	35-55	55.5-56.2

* The viscosities in Table 5-6 were determined by the falling-ball method in 20% solutions in acetone at 20°C. The viscosities in Table 5-7 were determined by the same procedure in 20% solutions in 90—10 acetone—alcohol at 25°C.

two manufacturers, as shown in Tables 5-6 and 5-7.[2,3]

Chemical Resistance. Cellulose acetate is not affected by weak organic acids; mineral, animal, and vegetable oils; and gasoline. It is not resistant to alkalis except in higher acetylations. Cellulose acetate is the most solvent-resistant of all cellulose derivatives.

Flammability. Cellulose acetate burns slowly.

Heat Stability. It is resistant to degradation at higher temperatures.

Solubility. The solubility of cellulose acetate is more limited in range of solvents than that of the other cellulose derivatives. The solvents most commonly used are shown in Table 5-9.

Chlorinated solvents are more useful for the higher acetylation products than for the low grades, but they are toxic and should be used with care.

Plasticizer Compatibility. Cellulose acetate has a much more limited range of plasticizers than the other cellulose derivatives. The most commonly used are:

Methox
Santiciser M-17
Santiciser E-15
Santiciser 8
Santiciser 9
Triphenyl Phosphate

Most of the other plasticizers for cellulose acetate have limitations of compatibility, water resistance, heat stability, or volatility.

For more complete data on cellulose acetate plasticizers reference can be made to the trade literature.*

The proportion of plasticizer used for paper coatings depends on the type of acetate, application, and type of paper to be coated but, in general it should be between 50 and 80% by weight, based on the acetate.

Resin Compatibility. Cellulose acetate has very limited compatibility with resins. Among those that are useful are the Acryloids, Santolites, some Rezyls, and some Durez resins.

General Compatibility. Cellulose acetate is incompatible with ethyl cellulose, chlorinated rubber, polyvinyl resins, acrylic resins, and polystyrene. It has only very limited compatibility with cellulose acetate–butyrate and cellulose acetate–propionate. It is compatible, however, with nitrocellulose.

---

* For example, "Hercules Cellulose Acetate"—Hercules Powder Co. "Eastman Cellulose Acetate"—Tennessee Eastman Co.

TABLE 5-8

General Properties of Cellulose Acetate [3]

Appearance	Flake or granules			
Color	None			
Odor	None			
Taste	None			
	% Acetic acid content			
	53	55	56	59
Specific gravity 25/4 °C	1.357	1.355	1.361	1.332
Refractive index	1.4780	1.4780	1.4770	1.4735
Penetration melting point, °C	194	188	200	202
Light transmission (0.002 in. film)	All solar radiation down to 2900 Å (resistant to discoloration by sunlight)			
Moisture absorption of flake, at equilibrium, at 25°C and 97-100% relative humidity	16.5	—	14.8	—

TABLE 5-9

Common Solvents and Diluents for Cellulose Acetate

Solvents	*Latent solvents*	*Diluents*
Acetone	Methyl Ethyl Ketone	Benzol
Methyl Acetate	Ethyl Acetate	Toluol
Methyl Cellosolve Acetate	Cellosolve Acetate	Petroleum Naphthas
Methyl Cellosolve	Ethyl Alcohol	
Nitromethane	Nitropropane	
Ethyl Lactate		
Diacetone Alcohol		

TABLE 5-10

Comparative Viscosities, LL-1 Type
Cellulose Acetate (9%) [3]

Solvent	*Comparative viscosity, cP at 25°C*
Acetone	46
Methyl Acetate	62
Methyl Cellosolve Acetate	350
Methyl Cellosolve	360
Nitromethane	207
Ethyl Lactate	720
Diacetone Alcohol	650

Mixtures	*Comparative viscosity, cP at 25°C*
9 Ethylene Dichloride +1 Ethanol	200
9 Acetone +1 Ethanol	42
9 Nitromethane +1 Ethanol	120
7 Nitromethane +3 Ethanol	102
9 Methyl Ethyl Ketone +1 Methanol	62
8 Methyl Ethyl Ketone +2 Methanol	70

Comparison of Cellulose Acetate with Nitrocellulose for Paper Coatings

Advantages	Disadvantages
Better heat stability	Higher cost
Better light stability	Much narrower range of solvents,
Much lower inflammability	plasticizers, and resins
Better resistance to grease	Higher water absorption
Easier to formulate for heat sealing	Higher moisture-vapor transmission
	Poorer adhesion

Cellulose Mixed Esters: Cellulose Acetate Butyrate and Cellulose Acetate Propionate

Chemical Constitution. Cellulose mixed esters are made by esterifying cellulose partly with acetic and partly with propionic or butyric acid and anhydrides. Like cellulose acetate, these mixed esters are usually partially hydrolized. Because both the ratio of acetyl to higher acyl content, and the degree of hydrolysis are variable, more types are theoretically possible than with the straight cellulose acetate. However, for the sake of commercial standardization the types available are very limited in number.

Cellulose tripropionate has a propionyl content of 51.8%, whereas cellulose tributyrate has a butyryl content of 57.3%. With most of the standard types, the free hydroxyl groups may vary from none to one in every two anhydroglucose units.

Effect of the Higher Acyl Content and the Degree of Hydrolysis on the Properties. The chemical and physical properties of the mixed esters depend on the amount of higher acyl group, the degree of hydrolysis,

TABLE 5-11

Types of Available Mixed Cellulose Esters[2]

Type	% Acetyl	% Propionyl	% Butyryl	Degree of hydrolysis *	Viscosity range, sec
AP-141-65	29-31	13-15	—	1	—
AP-452-20	1-3	43-47	—	2	—
AB-161-2	31.0	—	17.0	1	1-5
AB-161-20	31.0	—	17.0	1	10-30
AB-161-40	31.0	—	17.0	1	30-50
AB-172-15	30.0	—	17.5	2	10-16
EAB-381-20	13.0	—	37.0	1	15-30
AB-272-3	20.5	—	26.5	—	2-5
AB-272-20	20.5	—	26.5	—	10-40
AB-500-1	6.0	—	48.0	—	0.25-2.0
AB-500-5	6.0	—	48.0	—	3-8

* The figures in this column refer to the average number of free hydroxyl groups in every four anhydroglucose units.

and the viscosity. In general, increasing the higher acyl content results in increased moisture resistance, greater solubility range, better compatibility with resins and plasticizers, and greater electrical resistance. However, increasing the higher acyl content lowers the tensile strength and hardness. As hydrolysis increases, the tolerance for alcohols increases, whereas the tolerance for hydrocarbons and chlorinated solvents decreases. Tensile strength and flexibility increase with an increase of viscosity of the mixed esters.

Solubility. The solubility of the mixed esters depends on the higher acyl content and on the degree of hydrolysis. In general, the solubility range is much greater than that of cellulose acetate. Some of the most useful solvents are shown in Table 5-13.

TABLE 5-12

General Properties of Mixed Cellulose Esters

Appearance	Flake or granule
Color	Water-white
Odor	None
Specific gravity	1.21-1.25
Refractive index	1.48
Stability to boiling water (acidity developed in 8 hr)	0.10% maximum
Acidity (free acetic acid—ASTM method)	0.10% maximum
Melting range	Depends on type
Resistance to ultraviolet	Excellent
Burning rate	Low

TABLE 5-13

Typical Solvents for Mixed Cellulose Esters

Acetone
Methyl Ethyl Ketone
Methyl Butyl Ketone
Methyl Acetate, 88%
Ethyl Acetate, 88%
Butyl Acetate, 88%
Cellosolve Acetate
Methyl Cellosolve Acetate
Ethyl Lactate
1- and 2-Nitropropane
Diacetone Alcohol
Methylene Dichloride
Ethylene Dichloride
50% Toluol—50% Methyl Alcohol
50% Toluol—50% Ethyl Alcohol
80% Ethylene Dichloride—20% Ethyl Alcohol

Plasticizer Compatibility. The mixed esters are compatible with nearly all plasticizers, particularly those of the ester type. With the excep tion of the acetobutyrate types AB-500-1 and AB-500-5, they are incompatible with oils, either raw or treated, except in very small proportions when large quantities of plasticizers are used. The AB-500-1 and AB-500-5 types are compatible with linseed, castor, and soybean oils. The amount of plasticizer required for paper coatings is approximately 60—70%, based on the weight of the mixed ester.

Resin Compatibility. The compatibility of resins with the mixed esters is greater than with cellulose acetate, but not as great as with nitrocellulose. The resins listed in Table 5-14 are compatible, to a large degree, with the mixed esters.

TABLE 5-14

Typical Resins Compatible with Mixed Cellulose Esters

Abalyn	Resins A10, B7, B10
Amberol 226	(Resinous Products
Aroclor 1262, 4465,5460	& Chemical Co.)
Bakelite 3180, 4036	Rezyl 14
XR4503, BR254,	Santolites MH, MHP,
BR571B	M5
Gelva 2½	Teglac 15
Glycol Phthalate	Vinsol
Hercolyn	Durez 500, 525, 550, 209,
Cumar MS	175
	Paraplex G-25

General Compatibility. The mixed esters are compatible with each other, with nitrocellulose, and with polyvinyl acetate in some proportions. They are not compatible with cellulose acetate, ethyl cellulose, or the copolymer of vinyl chloride–vinyl acetate.

Comparison of Mixed Cellulose Esters with Nitrocellulose for Paper Coatings

Advantages	*Disadvantages*
Better heat stability	Higher cost
Better light stability	Narrower range of plasticizers and
Low inflammability	resins
	Higher moisture-vapor transmission
	Poorer adhesion

Comparison of Mixed Cellulose Esters with Cellulose Acetate for Paper Coatings

Advantages	*Disadvantages*
Better moisture resistance	Higher cost
Wider choice of solvents	Lower tensile strength
Greater compatibility with resins and plasticizers	

Ethyl Cellulose

Chemical Constitution. Ethyl cellulose is a cellulose ether made by the reaction of ethyl chloride with alkali cellulose. The completely-reacted product is the triethyl derivative, but the most useful compounds are those containing 2.15—2.58 ethoxy groups per anhydroglucose unit. This corresponds to an ethoxy content of 43.5—49.5%. In general, as the degree of substitution increases, the solubility in aromatic hydrocarbons increases and the melting point decreases. Lower degrees of substitution require larger amounts of alcohol in the solvent. Ethyl cellulose may be represented by the following structural formula:

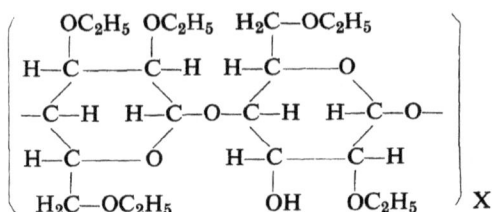

$$
\left(\begin{array}{c}
\text{OC}_2\text{H}_5 \quad \text{OC}_2\text{H}_5 \quad \text{H}_2\text{C}-\text{OC}_2\text{H}_5 \\
| \qquad\qquad | \qquad\qquad | \\
\text{H}-\text{C}\!-\!-\!-\text{C}-\text{H} \quad \text{H}-\text{C}\!-\!-\!-\text{O} \\
\diagup \qquad\qquad \diagdown \qquad\qquad \diagup \\
-\text{C}-\text{H} \quad \text{H}-\text{C}-\text{O}-\text{C}-\text{H} \quad \text{H}-\text{C}-\text{O}- \\
\diagdown \qquad\qquad \diagup \qquad\qquad \diagdown \\
\text{H}-\text{C}\!-\!-\!-\text{O} \quad \text{H}-\text{C}\!-\!-\!-\text{C}-\text{H} \\
| \qquad\qquad\qquad | \qquad\qquad | \\
\text{H}_2\text{C}-\text{OC}_2\text{H}_5 \qquad \text{OH} \quad \text{OC}_2\text{H}_5
\end{array}\right)_X
$$

Available Types of Ethyl Cellulose. The types shown in Tables 5-15 and 5-16 are listed by two manufacturers.

TABLE 5-15

Viscosities of Commercial Ethyl Celluloses [4]

48.5-50% Ethoxyl T	46.8-48.5% Ethoxyl N†	45.5-46.8% Ethoxyl K	44.5-45.5% Ethoxyl G	43.5-44.5% Ethoxyl D	*Viscosity of 5%, cP*	
—	x	—	—	—	6.5-7.5	viscosity of K, G, D
x*	x	—	—	—	9.5-10.5	types determined
x	x	x	—	—	13-15	in 70/30 toluene/
x	x	x	x	x	20-24	alcohol by weight;
x	x	x	x	x	42-55	viscosity of T, N
x	x	x	x	x	90-105	types determined
x	x	x	x	x	180-240	in 80/20 toluene/
x	x	—	—	—	270-330	alcohol by weight
—	x	—	—	—	330 & above	

* Denotes types commercially available.
† This type is the standard and is the most commonly used of the above for paper coatings.

The standard type of ethyl cellulose is the most generally useful type of this series, in that it has wider solubility and greater compatibility with plasticizers, resins, and waxes. However, the medium ethoxy type has a higher softening point and greater flexibility.

TABLE 5-16

Viscosities of Commercial Ethyl Celluloses [5]

Standard-Ethoxy Type Ethoxy Content 48.5-49.5%		Medium-Ethoxy Type Ethoxy Content 45.0-46.5%	
Type	*Limits*	*Type*	*Limits*
7	6-8	7	6-8
10	9-11	10	9-11
13	12-14	15	12-15
20	16-20	20	17-22
50	35-50	50	35-60
100	85-110	100	75-120
150	135-170	150	135-170
300	220-330		

TABLE 5-17

General Properties of Ethyl Cellulose (N Type) [4]

Appearance	Granular powder
Color	Water-white
Odor and taste	None
Specific gravity	1.14
Refractive index	1.47
Melting point	190-200°C
Softening point	140-150°C
Light transmission	Transparent down to 2800 Å (passes sunlight)
Effect of sunlight	Very slight
Moisture absorption (80% RH, 24 hr)	2.0%
Chemical Resistance	Fair resistance to weak acids; poor resistance to strong acids; excellent resistance to strong alkalis; good resistance to mineral oils, but poor resistance to animal and vegetable oils
Heat Stability	Retains flexibility after 72 hr at 100°C
Flammability	Low, about the same rate as paper

Solubility. Ethyl cellulose is soluble in a wide range of solvents, including esters, aromatic hydrocarbons, some alcohols, chlorinated solvents, and some ketones. Very useful solvent combinations can be made from mixtures of 70—90% aromatic hydrocarbons and 30—10% alcohol. The solvent combination used has a marked effect on the mechanical properties of the film deposited. Solvent mixtures of acetone and alcohol, for example, produce films of poor flexibility, whereas some mixtures of aromatic hydrocarbons and alcohols produce films of much better flexibility. Some of the most useful solvents are shown in Table 5-18.

TABLE 5-18

Common Solvents for Ethyl Cellulose

Esters	Alcohols	Ketones
Methyl Acetate	Methyl Alcohol	Acetone
Ethyl Acetate	Ethyl Alcohol	Methyl Ethyl Ketone
Isopropyl Acetate	Isopropyl Alcohol	
Butyl Acetate	Butyl Alcohol	
Amyl Acetate	Amyl Alcohol	
Methyl Cellosolve Acetate		
Cellosolve Acetate		
Ethyl Lactate		

Glycol Ethers	Hydrocarbons	Chlorinated Solvents
Cellosolve	Benzene	
Butyl Cellosolve	Toluene	Methylene Chloride
	Xylene	Ethylene Chloride

The effect of the solvent used on viscosity is shown in Table 5-19.

TABLE 5-19 [5]

Viscosity of Various Solutions of Dow Ethocel
(10 cP Type Ethocel—15 g in 100 cm^3 Solvent)

Solvent	Viscosity, cP at 25°C
80:20 Toluol—Ethanol	258
Methyl Ethyl Ketone	317
Ethyl Acetate, 99%	357
Ethylene Dichloride	466
Ethanol, Formula 30	560
Butyl Acetate, 90%	594
n-Butanol	1905
Toluol	1930

Plasticizer Compatibility. Ethyl cellulose is compatible with nearly all types of plasticizers and oils.

TABLE 5-20

Common Plasticizers Compatible with Ethyl Cellulose

Dibutyl Phthalate	Butyl Stearate
Diamyl Phthalate	Plasticizer 3GH
Methyl Cellosolve Phthalate	Dibutyl Sebacate
Tricresyl Phosphate	Nujol
Dow Plasticizers No. 2, 5, 6	Hercolyn
Santicizer B-16	Raw Castor Oil
Santicizer M-17	Baker's P-1 to P-16, inclusive
Methox	Baker's Pale Blown Castor Oils No.
Ethox	4, 16, 20, 1000, 75, 85, 2AC

Ethyl cellulose requires less plasticizer for a given elongation than nitrocellulose, cellulose acetate, or the mixed esters. For paper coatings the proportion required is about 25—35%, by weight, based on the ethyl cellulose.

Resin Compatibility. Ethyl cellulose is compatible with many resins, both natural and synthetic. These include many phenolics and some alkyds. The manufacturers' pamphlets[4,5] on ethyl cellulose should be consulted for a comprehensive list.

TABLE 5-21

Common Resins Compatible with Ethyl Cellulose

Dewaxed Dammer	Amberlac B-94
Dewaxed Shellac	Amberol 800
Elemi	Aroclor 5460
Wood Rosin	Bakelite XR2342
Staybelite A-1	Bakelite BR254
Super Beckacite 1001, 2000	Bakelite BR8900
Cumar P-10	Beckacite 1100, 1101, 1110
Durez 210, 550	Paraplex RG$_2$
Lewisol 1, 2, 18, 28, 150	Petrex No. 1
Nevillac, Soft	Rezyl 116
Nevillac, Hard	Santolite MHP, MS
Ester Gum	Teglac 2154
Vinsol	

Compatibility with Other Cellulose Derivatives. Ethyl cellulose is miscible with nitrocellulose in all proportions.

Comparison of Ethyl Cellulose with Nitrocellulose for Paper Coatings

Advantages	Disadvantages
Solubility in cheap solvents	Higher cost
Low inflammability	Poorer adhesion
Low discoloration by sunlight	More thermoplastic
Flexibility at low temperatures	Difficult to emboss without sticking
Heat resistance	Higher water absorption
Resistance to alkalis	Higher water-vapor permeability
Low specific gravity	Poorer resistance to animal and vegetable oils

Benzyl Cellulose

Benzyl cellulose is the benzyl ether of cellulose and is made by reacting alkali cellulose with benzyl chloride. It is much softer than the other cellulose derivatives and requires little plasticizer. It is soluble in aromatic hydrocarbons, esters, the higher ketones, and chlorinated hydrocarbons, but is insoluble in alcohols and ethers. It is compatible with many resins and most of the common plasticizers. It has a lower moisture absorption and lower moisture permeability than any of the other cellulose derivatives. Next to ethyl cellulose, it has the lowest specific gravity (1.230) of the cellulose derivatives. It has a lower tensile strength than the other cellulose derivatives. One of its outstanding disadvantages is its poor stability to ultraviolet light.

Benzyl cellulose is a comparatively new cellulose derivative and, has found little or no application to paper coating. For more detailed information about its properties, refer to the pamphlet "Benzyl Cellulose" from the Hercules Powder Company.

Polyvinyl Acetate

Chemical Constitution. Polyvinyl acetate is a thermoplastic resin made by the polymerization of the vinyl acetate monomer, which is by made the reaction of acetylene with acetic acid in the presence of a catalyst. Its chemical structure may be represented by the following formula:

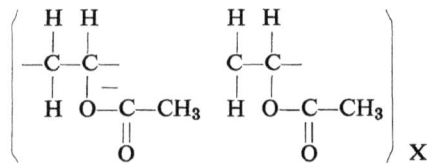

$$\left(\begin{array}{cc} \underset{\underset{\displaystyle H}{|}}{\overset{\overset{\displaystyle H}{|}}{C}} - \underset{\underset{\displaystyle O-C-CH_3}{|}}{\overset{\overset{\displaystyle H}{|}}{C}} - & \underset{\underset{\displaystyle H}{|}}{\overset{\overset{\displaystyle H}{|}}{C}} - \underset{\underset{\displaystyle O-C-CH_3}{|}}{\overset{\overset{\displaystyle H}{|}}{C}} - \\ \underset{\displaystyle O}{\|} & \underset{\displaystyle O}{\|} \end{array} \right)_X$$

Types Available. The only important variable in polyvinyl acetate is the degree of polymerization or molecular weight. The types available, therefore, differ only in this respect. Film strength, viscosity in solution, and softening temperature increase with increasing molecular weight. The series shown in Table 5-22 is representative of the commercially available types.[6]

TABLE 5-22

Properties of Some Commercially Available Polyvinyl Acetate Resins

Type	Weight % in acetone to give 250 cP viscosity at 20°C	Softening point, °C
AYAB	53	44.0
AYAA	39	66.0
AYAF	31	77.0
AYAT	26	86.5

The only types of practical importance for paper coating are the higher molecular weight resins AYAF and AYAT. The other grades are too soft and too thermoplastic to be used for nontacky coatings unless highly modified with other materials.

Solubility. Polyvinyl acetate is soluble in esters, ketones, chlorinated hydrocarbons, and lower aromatic hydrocarbons. It is also soluble in methanol, but not in the other lower-boiling alcohols unless they contain a small amount of water.

TABLE 5-23

General Properties of High Molecular Weight Polyvinyl Acetate Resins

Appearance	Granular white powder
Burning rate	Low
Effect of aging	Unaffected
Effect of sunlight	Unaffected
Heat stability at 135°C	No appreciable change after 4.5-5 hr
Water absorption at 25°C, 16 hr	1.4%
Specific gravity 20/20 °C	1.19
Refractive index	1.4665
Tensile strength	4200 lb/in.^{2}
Odor and taste	None
Chemical resistance	Not resistant to strong acids and alkalis. Resistant to dilute acids and alkalis and oils

Plasticizer Compatibility. Polyvinyl acetate is compatible with most of the common lacquer plasticizers. Some of the most useful plasticizers are listed in Table 5-24.

TABLE 5-24

Common Plasticizers for Polyvinyl Acetate Resins

Dibutyl Phthalate	Kronisol
Diamyl Phthalate	Methox
Dibutyl Sebacate	Santiciser B-16
Flexol 3GH	Santiciser M-17
Flexol 3GO	Triphenyl Phosphate
KP-120	

It is not compatible with castor oil, either raw or pale blown. Polyvinyl acetate requires much less plasticizer than nitrocellulose, the range being approximately 10—15% by weight, based on the resin.

TABLE 5-25

Common Resins Compatible with Polyvinyl Acetate

Acryloid C-10	Rezyl 14
Aroclor #1254	Santolites MS, MHP
Bakelite Phenolic Resin XR3180	Vinsol Resin
Super Beckacite Resin 1001	

Resin Compatibility. The representative resins listed in Table 5-25 are compatible with polyvinyl acetate.

General Compatibility. Polyvinyl acetate is compatible with nitrocellulose, ethyl cellulose, cellulose acetopropionate, cellulose acetobutyrate (in some proportions), and with polyvinyl butyral in the presence of mutual plasticizers. It is incompatible with cellulose acetate and with polyvinyl chloride.

General Remarks. Polyvinyl acetate, even in the higher molecular weights—in spite of its desirable properties of freedom from odor, taste, and toxicity, its good aging properties and sunlight resistance, and its wide solubility and compatibility—has serious drawbacks for paper coatings. It has very high solvent retention, which requires a forced dry bake at an elevated temperature for its removal. The dry film is very thermoplastic and has a high tendency to block. It finds specialized use for heat-seal coatings where close control between temperatures of blocking and heat seal is required. Unless there is a special reason for using polyvinyl acetate, the copolymer is much more suitable for paper coating.

Comparison of Polyvinyl Acetate with Nitrocellulose for Paper Coatings

Advantages	*Disadvantages*
Much lower specific gravity	Much higher solvent retention
Better sunlight resistance	Requires bake at elevated tempera-
Much slower burning	ture for maximum adhesion
Much easier to heat-seal	Exhibits some thixotropic behavior
Soluble in aromatic solvents	Not compatible with castor oil

Comparison of Polyvinyl Acetate with Copolymer Vinyl Chloride— Vinyl Acetate for Paper Coatings

Advantages	*Disadvantages*
Lower specific gravity	Poorer mechanical properties, i.e.,
Wider range of solubility	tensile strength, flexibility, and
Less thixotropic behavior	toughness
Easier to heat-seal	Poorer resistance to oils and greases
Compatibility with most of the	Poorer resistance to acids and alka-
cellulose derivatives	lis
	Greater tendency to block
	Poorer water resistance

Polyvinyl Chloride

Straight polyvinyl chloride has only limited application for paper coatings in a solvent medium. Its chief drawback is its limited solubility. It exhibits thixotropic behavior in almost any workable concentration and must be applied warm to obtain usable solid concentrations.

TABLE 5-26

Solvents for Polyvinyl Chloride

Acetonylacetone	Isophorone
Cyclohexanone	Methyl Isobutyl Ketone (Hot)
Dioxane	Methyl Ethyl Ketone (Hot)
Ethylene Dichloride	Tetrahydrofuran
Mesityl Oxide	

There are few applications for paper coating where the much greater solubility and better formulating properties of the copolymers do not make them preferable to the straight polyvinyl chloride.

Vinyl Chloride–Vinyl Acetate Copolymer

Chemical Constitution. The copolymers of vinyl chloride and vinyl acetate are made by copolymerizing these monomers in the presence of a catalyst. The vinyl acetate monomer can be made by the reaction of acetylene with acetic acid, whereas the vinyl chloride monomer is the reaction product of acetylene and hydrochloric acid in the presence of catalysts. The proportion that results in the most useful copolymers for coating is roughly 9 moles vinyl chloride to 1 mole vinyl acetate. This ratio produces the optimum combinations of the strength, toughness, and chemical resistance of polyvinyl chloride with the good solubility of polyvinyl acetate. The chemical structure may be represented as follows:

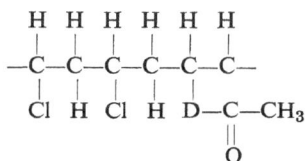

$$-\overset{\overset{\displaystyle H}{|}}{\underset{\underset{\displaystyle Cl}{|}}{C}}-\overset{\overset{\displaystyle H}{|}}{\underset{\underset{\displaystyle H}{|}}{C}}-\overset{\overset{\displaystyle H}{|}}{\underset{\underset{\displaystyle Cl}{|}}{C}}-\overset{\overset{\displaystyle H}{|}}{\underset{\underset{\displaystyle H}{|}}{C}}-\overset{\overset{\displaystyle H}{|}}{\underset{\underset{\displaystyle D-\overset{\overset{\displaystyle }{}}{\underset{\underset{\displaystyle O}{\|}}{C}}-CH_3}{|}}}{C}}-\overset{\overset{\displaystyle H}{|}}{C}-$$

Types Available. Both the proportion of vinyl chloride to vinyl acetate, and the viscosity of the resulting polymer can be varied, so that several types are possible. As the chain length of the copolymer increases, the solubility is reduced, whereas the solution viscosity, the softening point, and film strength increase. Table 5-27 lists the types of this copolymer produced by one manufacturer.[7]

The most suitable type for paper coatings is probably the VYHH-1 resin. It represents a good compromise between the low-viscosity, highly soluble, and less flexible, lower molecular weight resin VYLF and the higher-viscosity, less soluble, tougher, and more flexible higher molecular weight resin VYNS-3. The type represented by VYDR is

TABLE 5-27
Properties of Various Polyvinyl Chloride—Acetate Resins

Type	Approximate Chemical Composition, % by wt.			Intrinsic Viscosity (in Cyclo-hexanone at 20°C)	Specific Gravity	Solvents Commonly Used
	V.C.	V.A.	Other			
VYCC	62	38	—	0.28	1.30	
VYLF	87	13	—	0.24	1.34	Ketones, selected esters, tolerates aromatic hydro-carbons
VYHH-1	87	13	—	0.53	1.36	
VMCH	86	13	1	0.53	1.35	
VAGH	91	3	6.0*	0.57	1.39	
VYNS-3	90	10	—	0.79	1.36	Ketones
VYDR	95	5	—	1.25	1.39	Cyclic ketones at room temperature

* Polyvinyl alcohol.

of high viscosity, tough, elastic, and nontacky, but its restricted solubility limits its use to special paper coatings.

VMCH is a special type with approximately 1 % of interpolymerized maleic acid. This small amount of dibasic acid gives it improved adhesion to many surfaces.

Chemical Resistance. The copolymers are extremely resistant to weak and strong acids, to weak and strong alkalis (except ammonia), and to oxidizing and reducing agents. They are not affected by alcohol, grease, or petroleum hydrocarbons and are permanently thermoplastic and nonoxidizing.

Solubility. The solubility of the copolymers depends greatly on the composition and molecular weight. The lower molecular weight VYLF resin is much more soluble than the higher molecular weight VYHH-1, which in turn is much more soluble than the still higher molecular weight resins VYNS-3. Because the VYHH-1 type is the most widely used for paper coatings, the solubility discussion is limited to this grade.

The copolymer VYHH is readily dissolved at room temperature by ketones, esters, nitroparaffins, and some chlorinated hydrocarbons. Aromatic hydrocarbons are not solvents, but can be used to a great extent as diluents. The copolymer is highly insoluble in alcohols and

TABLE 5-28

General Properties of a Commercial Polyvinyl Chloride— Acetate Resin (VYHH)

Appearance	White powder
Specific gravity	1.36
Index of refraction	1.525
Flammability	Will not support combustion
Odor	None
Taste	None
Toxicity	None
Heat stability	Poor, above 300°F; has a tendency to split off HCl; must be stabilized
Light stability	Relatively poor light stability in clear films unless properly stabilized
Adhesion	Poor when air dried; should be baked for short time in excess of 300°F for maximum adhesion

TABLE 5-29

Common Solvents for Polyvinyl Chloride—Acetate Copolymer Resin

Acetone	Isopropyl Acetate (98%)
Methyl Ethyl Ketone	Methyl Cellosolve
Di-Isobutyl Ketone	Methyl Cellosolve Acetate
Methyl n-Amyl Ketone	Dioxane
Methyl Isobutyl Ketone	Cyclohexanone
Tetrahydrofuran	Ethylene Dichloride
Ethyl Acetate (98%)	n-Butyl Acetate (98%)

aliphatic hyprocarbons.

In general, the ketones are the most suitable solvents. They permit higher concentrations without gelling, lower solution viscosities for the same solid content, and the use of greater amounts of diluents.

In considering the solubility of the copolymer, certain solution characteristics which are not generally encountered in solutions of the cellulose derivatives, must be kept in mind.

In a given solvent or solvent combination, normal increase in viscosity results with increased resin content up to a certain point. Above this point, the viscosity increases much more rapidly with concentration, and is much more affected by slight changes in temperature and agitation. At this stage, the solution exhibits marked thixotropic behavior. If the concentration is increased still further above this point, a stage is reached at which the solution gels. For each solvent combination, a phase diagram can be drawn showing the concentration area for these three types of solubility behavior. Changes in the amount of diluent often produce marked changes in the phase diagram characteristics.

Plasticizers. In general, about 20—25% plasticizer, based on the amount of the copolymer, is the optimum proportion for paper coatings. The most commonly used plasticizers are listed in Table 5-30.

TABLE 5-30

Plasticizers for Polyvinyl Chloride—Acetate Copolymers

Dioctyl Phthalate
Dicapryl Phthalate
Butoxyglycol Phthalate (Kronisol)
Methoxyglycol Acetyl Ricinoleate (KP-120)
Tricresyl Phosphate
Methyl Phthalyl Methyl Glycollate (M-17)
Butyl Phthalyl Butyl Glycollate (B-16)
Ethyl Phthalyl Ethyl Glycollate (E-15)
Ethoxydiglycol Phthalate
Ethoxyglycol Phthalate (Ethox)
Methoxyglycol Phthalate (Methox)
Acetylated Castor Oil (P-8)
Butyl Ester of Acetylated Polymerized Ricinoleic Acids (P-16)
Chlorinated Paraffin
Plasticizer S.C.
Paraplex G-25

Resin Compatibility. The compatibility of the copolymers VYLF, VYHH-1, VMCH, VYNS, and VDYR with the usual lacquer resins is quite limited. As a rule, little or no advantage is gained by the use of resins with these copolymers for paper coatings. Because of the limited compatibility, two new types, VYCC and VAGH, which overcome this limitation to a great extent, have appeared on the market. VYCC is compatible with nitrocellulose in all proportions. VAGH is compatible with most synthetic resins. Some of the resins compatible with the older grades are shown in Table 5-31.

General Compatibility. The polyvinyl chloride-acetate copolymer is incompatible with nitrocellulose, cellulose acetate, cellulose acetobuty-

TABLE 5-31

Common Resins Compatible with Polyvinyl Chloride—Acetate Copolymers

Acryloid A10, B-7
Aroclor 5460
Bakelite Resin BR254, BR3360, BR4036
Durez Phenolic Resin 550
Methyl, Ethyl, n-Propyl, n-Butyl, Isobutyl Methacrylates
Santolite MS, MHP
Super Beckacite 1001 and 2000
Rosin W W

rate, cellulose acetopropionate, ethyl cellulose, polyvinyl acetate, and polyvinyl butyral.

Comparison of Polyvinyl Chloride—Acetate with Nitrocellulose for Paper Coatings

Advantages	*Disadvantages*
Tougher	Narrower range of useful solvents, plasticizers, and resins
More flexible	
Better resistance to strong acids and alkalis	Exhibits thixotropic behavior and gel formation above certain concentrations
Better resistance to alcohols, greases, and oils	Much higher solvent retention
Noninflammability	Requires bake at elevated temperature for best adhesion
Easier to heat-seal	
Better sunlight resistance of stabilized coatings	Poor heat stability at baking temperatures unless properly stabilized

Polyvinyl Butyral

Chemical Constitution. Polyvinyl butyral is made from polyvinyl acetate by a combination of hydrolysis and partial condensation with butyraldehyde. The commercially available polymer contains a small proportion of free hydroxyl groups, which are chemically reactive and can be utilized to render the polyvinyl butyral thermosetting by the addition of certain other reactive resins. Lowering the free hydroxyl content gives the polyvinyl butyral greater tolerance for hydrocarbons, lower viscosity, less chemical reactivity, and better water resistance of the uncured film. Polyvinyl butyral may be represented by the following structural formula; free hydroxyl groups are not shown:

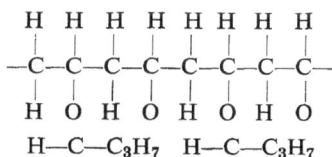

$$
\begin{array}{cccccccc}
H & H & H & H & H & H & H & H \\
| & | & | & | & | & | & | & | \\
-C\!-\!C\!-\!C\!-\!C\!-\!C\!-\!C\!-\!C\!-\!C- \\
| & | & | & | & | & | & | & | \\
H & O & H & O & H & O & H & O
\end{array}
$$

$$H\!-\!C\!-\!C_3H_7 \quad H\!-\!C\!-\!C_3H_7$$

Types Available. Table 5-32 lists the types of polyvinyl butyral resin marketed by one manufacturer.[8]

TABLE 5-32

Commercially Available Types of Polyvinyl Butyral Resin

Type	% Free hydroxyl	Intrinsic viscosity in cyclohexanone at 20°C	Specific gravity
XYHL	7	0.81	1.03
XYSG	7	1.16	1.03

TABLE 5-33

General Properties of Polyvinyl Butyral Resins

Appearance	White powder
Tensile strength	8,100-8,500 lb/in^2
Water absorption (16 hr at 60°C)	3-5%
Resistance to alkalis	Moderately good
Resistance to acids	Poor

Solubility. Polyvinyl butyral is soluble in methanol, dioxane, and glycol ethers. It is also soluble in ethyl, isopropyl, and butyl alcohols when these contain the normal amount of water present in their respective constant-boiling mixtures. It is not soluble in anhydrous alcohols except anhydrous methanol. It is also soluble in ester solvents when these are used with alcohols. Ketones, aromatic and chlorinated hydrocarbons swell the butyral, but do not dissolve it. Its tolerance for aliphatic hydrocarbons is poor.

Plasticizer Compatibility. The phthalates and phosphates, in general, are good solvent-type plasticizers for polyvinyl butyral. Castor oil, blown soybean oil, most of the ricinoleates, blown linseed oil, and oleic acid are also useful as plasticizers for these resins.

Resin Compatibility.[9] Among resins compatible with polyvinyl butyral in some proportion are those shown in Table 5-34.

TABLE 5-34

Common Resins Compatible with Polyvinyl Butyral

Amberol # 226	Dewaxed Dammar
Bakelite BR-8900	Gum Elemi
Bakelite BV-1600	Hydrogenated Rosin
Bakelite XR-9432	Mastic
Bakelite BV-9700	Neville Hard
Beckacite 1003	Santolite MS
Beckasol B-1324	Santolite K
Durez # 550	Shellac
Congo Hard	

Comparison of Polyvinyl Butyral with Vinyl Chloride—Vinyl Acetate
Copolymer for Paper Coatings

Advantages	*Disadvantages*
Freedom from thixotropic and gel behavior	Higher cost
Castor oil can be used as a plasticizer in high proportion	Poorer water resistance
Can be reacted to form a nonthermoplastic coating	Poorer alkali resistance
Has greater general compatibility with resins and plasticizers	Poorer grease resistance
	Higher moisture-vapor transmission

General Compatibility. Polyvinyl butyral is compatible with nitrocellulose, but not with the other cellulose derivatives. It is incompatible with the vinyl chloride-vinyl acetate copolymer.

Vinyl Chloride–Vinylidene Chloride Copolymer

Chemical Consitution. The vinyl chloride monomer can be made by reacting acetylene with hydrochloric acid in the presence of catalysts. The vinylidene chloride is made from trichloroethane, which in turn stems from the reaction of chlorine with vinyl chloride. The copolymer described here is mostly vinyl chloride with a small proportion of vinylidene chloride. Its chemical structure may be represented as follows:

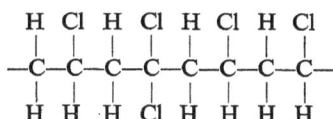

$$
\begin{array}{c}
\text{H} \quad \text{Cl} \quad \text{H} \quad \text{Cl} \quad \text{H} \quad \text{Cl} \quad \text{H} \quad \text{Cl} \\
| \quad | \quad | \quad | \quad | \quad | \quad | \quad | \\
-\text{C}-\text{C}-\text{C}-\text{C}-\text{C}-\text{C}-\text{C}-\text{C}- \\
| \quad | \quad | \quad | \quad | \quad | \quad | \quad | \\
\text{H} \quad \text{H} \quad \text{H} \quad \text{Cl} \quad \text{H} \quad \text{H} \quad \text{H} \quad \text{H}
\end{array}
$$

Types Available. Several commercial types of this copolymer are listed by one manufacturer.[10] The concentrations shown in Table 5-35 represent solutions of approximately 500 cP viscosity in methyl ethyl ketone, after aging at room temperature for 7 days. See Table 5-36 for general properties.

TABLE 5-35

Solution Viscosities of Commercial Vinyl Chloride-Vinylidene Chloride Copolymers (% solubilized)

Type	Dissolved at 30°C Viscosity at 30°C	Dissolved at 70°C Viscosity at 30°C	Dissolved at 70°C Viscosity at 70°C
Geon 202	7½	10	10
Geon 203	—	—	17½
Geon 204	15	20	20
Geon 205	17½	20	22½
Geon 200X14	30	35	35

TABLE 5-36

General Properties of Vinyl Chloride—Vinylidene Chloride Copolymers

Appearance	White powder
Odor	None
Taste	None
Toxicity	None
Specific gravity	1.43 (average)
Chemical resistance	Highly resistant to strong acids and alkalis and to greases and oils
Burning rate	Very low

Solubility. The copolymers have a narrow choice of solvents.

Methyl ethyl ketone is the most commonly used. Cyclohexanone and other higher ketones can be used with methyl ethyl ketone to obtain slightly higher concentrations. The solutions exhibit marked thixotropic and gel behavior. The best procedure for making solutions, particularly of the higher molecular weight grades, is to dissolve them at around 70°C. The solutions should be used above room temperature to avoid excessive gel formation.

Plasticizer Compatibility. Among the most useful plasticizers are those shown in Table 5-37.

TABLE 5-37

Common Plasticizers for Vinyl Chloride—Vinylidene Chloride Copolymers

Dioctyl Phthalate	Santicizer B-16
Dicapryl Phthalate	Plasticizer SC
Butoxy Ethyl Phthalate	

Resin Compatibility. Little information is available on resin compatibility, but the copolymer is generally compatible with acrylic resins.

Comparison of Vinyl Chloride—Vinylidene Chloride Copolymers with Nitrocellulose for Paper Coatings

Advantages	Disadvantages
Much tougher, more flexible, higher abrasive resistance	Higher cost
Better resistance to strong acids and alkalis, alcohols, greases, and oils	Much narrower range of solubility and compatibility with resins and plasticizers
Can be made nonflammable	Exhibits marked thixotropic and gel behavior
Easier to heat-seal	Much higher solvent retention
	Requires bake at elevated temperature for best adhesion
	Poor heat stability at baking temperatures, unless properly stabilized

Vinylidene Chloride–Acrylonitrile Resins

Some resins consist of copolymers of vinylidene chloride and acrylonitrile.

Chemical Constitution. Vinylidene chloride is made by the reaction of ethylene and chlorine in the presence of a catalyst. Acrylonitrile is made by the reaction of ethylene oxide and hydrogen cyanide. The

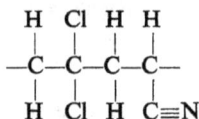

copolymer may be represented by the following formula:

Types Available. The general type described here is known commercially as Saran F-120 and is available in several viscosities, as listed
in Table 5-38.[11]

TABLE 5-38

Viscosity Range of Various Saran F-120 Resins

20% by weight solution in methyl ethyl ketone, cP	Range, cP
40	30-50
200	180-220
1,000	1,000-1,200
8,000	7,000-9,000

As is the case with all linear polymers, the tensile strength, toughness, flexibility, and solution viscosity increase with increasing molecular weight. The 200 and 1,000 cP types are the most suitable for paper
coating. For properties, see Table 5-39.

TABLE 5-39

General Properties of Vinyl Chloride—Acrylonitrile Resins

Appearance	White powder
Odor and taste	None
Toxicity	None
Specific gravity 25/4°C	1.60
Refractive index 25°C	1.580
Tensile strength	8-10,000 lb/in.^{2}
Elongation	8%
Moisture-vapor Transmission of 1 mil Film (g/100 in.^{2}/24 hr at 100°F and 90% RH)	0.15
Flammability	None
Chemical resistance	Resistant to weak and strong acids, weak alkalis; unaffected by oils and grease

Solubility. The most satisfactory solvent for this copolymer
is methyl ethyl ketone. The only other readily available complete solvents are cyclohexanone, isophorone, and mesityl oxide. The mesityl
oxide solution gradually turns dark brown in storage. Esters, ketones,
chlorinated hydrocarbons, and aromatic hydrocarbons can be added
in varying proportions to replace part of the methyl ethyl ketone.
Table 5-40 shows the maximum percentage of methyl ethyl ketone
that can be replaced by diluents in a 15% solution of the 1,000 cP
copolymer.[11]

One of the great differences between the copolymer of vinylidene

TABLE 5-40

Maximum Percentage of Methyl Ethyl Ketone Replaceable by Diluents in a Standard Solution of Vinyl Chloride—Acrylonitrile Resin

	%		%
Acetone	75	Toluene	30
Methyl Isobutyl Ketone	90	Xylene	20
Methyl-n-Amyl Ketone	90	Methanol	25
Ethyl Acetate	60	Isopropyl Alcohol	15
Butyl Acetate	50	Cellosolve	15
Amyl Acetate	50	Ethylene Dichloride	35
Nitromethane	25	Trichlorethylene	40
Nitroethane	25	Propylene Oxide	60
2-Nitropropane	40	Dioxane	90

chloride and acrylonitrile and the copolymer of vinyl chloride and vinyl acetate is that the first resin has no tendency to form gels even at high solids concentrations and viscosities. Like other vinyl resins, however, it has high solvent retention that requires drying at elevated temperatures for thorough removal.

Plasticizer Compatibility. The copolymer requires little plasticizer. The optimum range is 0—15% by weight, based on the copolymer. The choice of solvent-type plasticizers is quite limited. Nonsolvent plasticizers can also be used in small amounts, and are helpful in decreasing the tack and improving block resistance.

Some of the representative plasticizers are listed in Table 5-41.

TABLE 5-41

Common Plasticizers Used with Vinyl Chloride—Acrylonitrile Copolymers

Solvent type	Nonsolvent type
Dibutyl Phthalate	Tricresyl Phosphate
Santicizer B-16	Triphenyl Phosphate
Kronisol	KP 23
Dibenzyl Sebacate	Butyl Stearate
Diphenyl Phthalate	KP 120
Ethox	Flexol 3 GO

In general, addition of plasticizer has a marked effect in increasing the moisture-vapor transmission of the plasticized film. This effect depends on the plasticizer and on the amount used.

TABLE 5-42

Resins Used with Vinyl Chloride—Acrylonitrile Copolymer Resins

Acryloid B-72	Paraplex G-25, AP 11
Ester Gum	Petrex A5, HT
Aroclors	Santolite MS
Lucite	Super Beckacite 2000

Resin Compatibility. Although most resins increase the moisture-vapor loss of the film,[11] some of them help to increase the gloss and decrease the solvent retention. Among the resins compatible in some proportions are those shown in Table 5-42.

General Compatibility. The copolymer has only very limited compatibility with nitrocellulose, vinyl acetate, vinyl chloride–vinyl acetate copolymer, and polyvinyl butyral.

Comparison of Vinyl Chloride Acrylonitrile Copolymers with Nitrocellulose for Paper Coatings

Advantages	Disadvantages
Tougher	Higher cost
More flexible	Greater solvent retention
Better moisture-vapor resistance	More tendency to block
Better resistance to oil and grease	Much more limited choice of solvents, plasticizers and resins

Comparison of Vinyl Chloride—Acrylonitrile Resins with Vinyl Chloride—Vinyl Acetate Copolymer for Paper Coatings

Advantages	Disadvantages
Better moisture-vapor resistance	Higher cost
No tendency for gel formation	Higher density
	More limited solvent choice
	Poorer light stability
	Greater solvent retention

TABLE 5-43

Commercial Types of Polyethylene Resin

Type	Melt viscosity at 130°C, cP	Specific gravity
DYLT	4,000	0.92
DYNH	1,000,000	0.92

Polyethylene

Chemical Constitution. Polyethylene is made by the polymerization of ethylene in the presence of a catalyst at medium to high pressures. It is a linear thermoplastic polymer, crystalline in structure, and may be represented by the following structural formula:

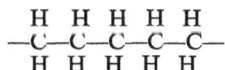

$$-\overset{\displaystyle H}{\underset{\displaystyle H}{C}}-\overset{\displaystyle H}{\underset{\displaystyle H}{C}}-\overset{\displaystyle H}{\underset{\displaystyle H}{C}}-\overset{\displaystyle H}{\underset{\displaystyle H}{C}}-\overset{\displaystyle H}{\underset{\displaystyle H}{C}}-$$

Types Available. By varying the conditions of polymerization, polyethylene can be made in many molecular weights. Lower molecular weight polymers are greaselike at room temperature, the high molecular weight polymers are tough, rubbery materials. The commercially available types, however, are very limited. The types shown in Table 5-43 are the only ones made in limited quantities.[11] See Table 5-44 for general properties.

TABLE 5-44

General Properties of Commercial Polyethylene Resin[12]

Appearance	White translucent granules
Odor	None
Taste	None
Toxicity	None
Tensile strength	1825 lb/in.^{2}
Softening temperature	108-112°C
Water absorption, 24 hr at 25°C	0.01%
Flexibility	Similar to that of copolymer VYNW plasticized with 20% dioctyl phthalate
Flammability	Slow burning
Chemical resistance	Good resistance to strong acids and alkalis and to all types of greases and oils at room temperature

Solubility. Polyethylene is insoluble in all organic solvents at room temperature. It is soluble in aromatic and chlorinated hydrocarbons at 180°F and must be used in hot solutions.

Plasticizer Compatibility. Polyethylene requires no plasticizer—a great advantage that assures freedom from migration.

General Remarks. Polyethylene has better solvent release than any of the vinyls. Temperatures of about 300°F are recommended for the final stages of drying.

Comparison of Polyethylene with Vinyl Chloride—Vinyl Acetate Copolymer for Paper Coatings

Advantages	Disadvantages
Much lower specific gravity	Hot solvent must be used
Cheaper coatings are possible	More inflammable
Wide range of flexibility from 90—230°F	Cannot be heat-sealed with electronic equipment
Unaffected by any organic solvent at room temperature	
No plasticizer required, which eliminates possibility of migration	
Less solvent retention	

Chlorinated Rubber

Chemical Constitution. Chlorinated rubber is made by treating pure crepe rubber dissolved in a solvent, e.g., carbon tetrachloride, with chlorine. The commercially available material contains approximately 67% chlorine. This is a mixture of two products, one of which has a chlorine content of 61.3%, and the other a chlorine content of 68.2%. These two chlorinated products may be represented as follows:

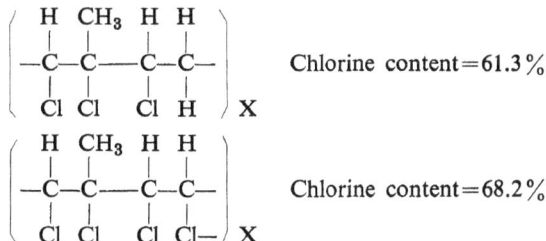

$$
\left(\begin{array}{cccc}
\overset{\displaystyle H}{|} & \overset{\displaystyle CH_3}{|} & \overset{\displaystyle H}{|} & \overset{\displaystyle H}{|} \\
-C- & C- & -C- & C- \\
\underset{\displaystyle Cl}{|} & \underset{\displaystyle Cl}{|} & \underset{\displaystyle Cl}{|} & \underset{\displaystyle H}{|}
\end{array}\right)_X \qquad \text{Chlorine content} = 61.3\%
$$

$$
\left(\begin{array}{cccc}
\overset{\displaystyle H}{|} & \overset{\displaystyle CH_3}{|} & \overset{\displaystyle H}{|} & \overset{\displaystyle H}{|} \\
-C- & C- & -C- & C- \\
\underset{\displaystyle Cl}{|} & \underset{\displaystyle Cl}{|} & \underset{\displaystyle Cl}{|} & \underset{\displaystyle Cl-}{|}
\end{array}\right)_X \qquad \text{Chlorine content} = 68.2\%
$$

Types Available. The types shown in Table 5-45 are commercially available.[13] See also Table 5-46.

TABLE 5-45

Commercially Available Types of Chlorinated Rubber

Viscosity (20% by weight in toluene at 20°C), cP	Viscosity type, cP
5—7	5
8—12	10
16—25	20
110—190	125
800—1,000	1,000

TABLE 5-46

General Properties of Chlorinated Rubbers [13]

Appearance	White granular powder
Color	Water-white
Odor and taste	None
Specific gravity	1.64
Index of refraction	1.554
Burning rate	Noninflammable
Effect of heat	Stable up to 125°C, decomposes at 135—150°C
Effect of sunlight	Discolors and embrittles
Effect of hot water	Blushes
Effect of cold water	None
Moisture absorption (80% RH for 24hrs)	0.27%
Chemical resistance	Excellent resistance to weak and strong acids and alkalis, good resistance to mineral oil, but poor resistance to animal and vegetable oils

For paper coatings, the most suitable viscosity types are the 125 cP and the 1,000 cP. The lower viscosity types are too brittle for coatings on flexible surfaces.

Solubility. Chlorinated rubber is soluble in nearly all types of common organic solvents except alcohols and aliphatic hydrocarbons. Other common solvents are given in Table 5-47.

TABLE 5-47

Common Solvents for Chlorinated Rubber

Ethyl Acetate	Ethylene Dichloride
Butyl Acetate	Benzol
Amyl Acetate	Toluol
Methyl Ethyl Ketone	Xylol
Methyl Isobutyl Ketone	Mixed Aromatic and Aliphatic
Cellosolve	Hydrocarbons
Diacetone Alcohol	

Acetone by itself is only a partial solvent, but mixtures with petroleum naphthas, which are themselves nonsolvents, are good solvent combinations.

Plasticizers. Chlorinated rubber is compatible with a great number of the common plasticizers, including those shown in Table 5-48.

TABLE 5-48

Common Plasticizers for Chlorinated Rubber

Dibutyl Phthalate	Santicizer B-16
Diamyl Phthalate	Soybean Oil
Cellosolve Phthalate	Tributyl Phosphate
Raw Castor Oil	Tricresyl Phosphate
Castor Oil Fatty Acid Esters	Triphenyl Phosphate
Butyl Stearate	Hercolyn
Camphor	Aroclor 1254, 1260
Santicizer M-17	

Chlorinated rubber usually requires 20—25 parts of plasticizer to 100 parts of chlorinated rubber, by weight, for paper coatings.

Compatibility with Resins. Chlorinated rubber is compatible with a great number of resins. With regard to resin compatibility, one manu-

TABLE 5-49

Resins Compatible with Chlorinated Rubber

Amberol ST-137	Neville R-3
Aroclor 5460	Rezyl 8075
Bakelite BR-1329	Staybelite
Bakelite BR-2963	Staybelite Ester Gum
Bakelite KJ-9868	Santolite MHP
Beckopol	Syntex 28
Cumar P10	Syntex 32
Dammar	Teglac 15
Ester Gum	Teglac Z-152
Ethyl Methacrylate	Vinsol

facturer states that of 137 resin-chlorinated rubber combinations, 87 showed good miscibility in both solution and film;13, slight immiscibility; and 36, pronounced immiscibility.[13] See Table 5-49.

Compatibility with Other Substances. Chlorinated rubber is also compatible with crude beeswax, paraffin wax, white ceresin wax in certain proportions, and chlorinated paraffin. It is not compatible with any of the cellulose derivatives or with vinyl acetate-vinyl chloride copolymer.

Application of Chlorinated Rubber to Paper Coatings

Advantages	Disadvantages
Noninflammable	Higher cost than nitrocellulose
Soluble in cheap solvents	Poorer strength and flexibility than nitrocellulose
High resistance to acids and alkalis	
Good moisture-vapor resistance	Poor sunlight resistance in clear films
Freedom from taste and odor	More thermoplastic than nitrocellulose, and therefore more tendency to block
	Poor resistance to animal and vegetable oils
	More solvent retention than nitrocellulose

Cyclized Rubber

Chemical Constitution. Cyclized rubber is made from low protein content natural crepe rubber, and has the same empirical formula as the rubber hydrocarbon. Crepe rubber is dissolved in a solvent, and the solution is refluxed in a reactor in the presence of a catalyst such as stannic chloride or chlorostannic acid. In the course of the reaction, some of the double bonds of the isoprene unite to form cross linkages. The time and conditions of the reaction determine, to a great extent, the physical characteristics of the resulting material. When the reaction is complete, the product, which has now become a thermoplastic resin far different from the original rubber, is precipitated from the solution and dried as a powder.

TABLE 5-50

Commercially Available Types of Cyclized Rubber

Type	Softening point, °C	Form
Normal Resin	55—65	Milled
P-843	25—30	Milled
P-704	40—47	Unmilled
P-744	40—47	Milled
P-40	75	Unmilled
P-20	105	Unmilled

The relation between isoprene, which is the chemical unit of natural rubber, and cyclized rubber may be represented as follows:

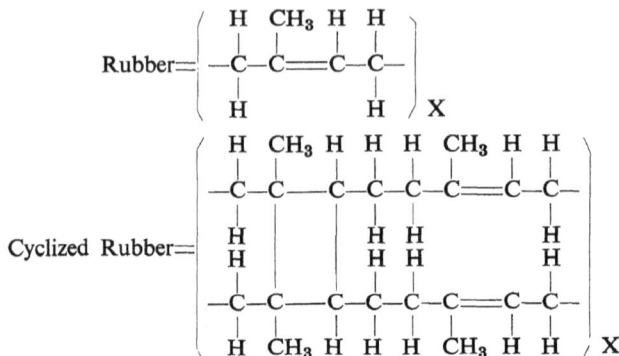

$$
\text{Rubber} = \left(
\begin{array}{c}
\text{H} \quad \text{CH}_3 \ \text{H} \quad \text{H} \\
| \qquad | \qquad | \qquad | \\
-\text{C}-\text{C}=\text{C}-\text{C}- \\
| \qquad\qquad\qquad | \\
\text{H} \qquad\qquad \text{H}
\end{array}
\right)_X
$$

$$
\text{Cyclized Rubber} = \left(
\begin{array}{c}
\text{H} \ \text{CH}_3 \ \text{H} \ \text{H} \ \text{H} \ \text{CH}_3 \ \text{H} \ \text{H} \\
-\text{C}-\text{C}-\!\!-\!\!-\text{C}-\text{C}-\text{C}-\text{C}=\text{C}-\text{C}- \\
\text{H} \qquad\quad \text{H} \ \text{H} \qquad\quad \text{H} \\
\text{H} \qquad\quad \text{H} \ \text{H} \qquad\quad \text{H} \\
-\text{C}-\text{C}-\!\!-\!\!-\text{C}-\text{C}-\text{C}-\text{C}=\text{C}-\text{C}- \\
\text{H} \ \text{CH}_3 \ \text{H} \ \text{H} \ \text{H} \ \text{CH}_3 \ \text{H} \ \text{H}
\end{array}
\right)_X
$$

According to these structural formulas, one half of the original double bonds remains unsaturated. To reduce the tendency for oxidation at these double bonds, use of an antioxidant is advisable.

Types Available. Both the viscosity and softening point of the cyclized rubber can be varied by varying the conditions of manufacture. The types shown in Table 5-50 are currently available under the trade name of Pliolite.[14]

Each of these resins is available in one of two forms: as milled or unmilled powder. Milling reduces solution viscosity and makes the resin easier to dissolve, because the unmilled powder tends to bunch up in solution and form lumps that are hard to break up. The milled resin, being in the form of pellets rather than powder, tends to oxidize less rapidly before it is put into solution.

In addition to the types of cyclized rubber made from natural rubber, substitute types from other rubberlike polymers were made during World War II. One of these was Pliolite S-1, a cyclized polyisoprene no longer being made.

A new type, known as Pliolite S-7, and which is made from synthetic rubber of the butadiene–styrene type, is in the final stages of development. It has good heat-seal and coating properties, and is considerably cheaper than the natural cyclized rubber.

The type most generally used for solution work is the normal resin with a melting point of 55 to 65°C.

Solubility. Cyclized natural rubber is soluble in petroleum, aromatic, terpene, and chlorinated hydrocarbons. It is not soluble in esters, ketones, or alcohols. The viscosity for a given concentration varies with the solvent used, as shown in Table 5-51.[15]

TABLE 5-51

Relative Viscosity of a 20% Solution of Cyclized Rubber in Various Solvents

Solvent	No. 4 Ford cup viscosity 73-75°F
Benzol	1.70
Toluol	1.395
Xylol	1.68
Solvesso # 2	1.925
Solvesso # 3	5.02
Turpentine	6.0

Plasticizer Compatibility. Cyclized rubber does not require much plasticizer. Based on the weight of the cyclized rubber, 5—10% is all that is needed for most applications. See Table 5-52 for common plasticizer.

TABLE 5-52

Common Plasticizers for Cyclized Rubber

Raw Tung Oil	Paraffin Wax
Castor Oil	Chlorinated Diphenyls
Butyl Stearate	Modified Glycerol Phthalate
Tricresyl Phosphate	

TABLE 5-53

Resins Compatible with Cyclized Rubber

Cumar Resins	Maleic Resins
Polyterpene Resins	Rosin
Modified Styrene Resins	Rosin Esters
Modified Phenolics	Dammar
Alkyds	

TABLE 5-54

General Properties of Cyclized Rubber

Specific gravity	1.07
Color	Slightly yellowish
Odor	None
Taste	None
Toxicity	None
Water resistance	Very good
Water-vapor resistance	Outstanding
Heat-seal properties	Very good
Chemical resistance	Resistant to weak and strong acids and alkalis, and alcohol. Poor resistance to gasoline and mineral, vegetable, and animal oils

Paraffin wax is probably the most useful of these plasticizers, because it not only plasticizes the film, but also greatly improves its water-vapor resistance.

Resin Compatibility. Cyclized rubber is compatible with the resins shown in Table 5-53.

For most applications, cyclized rubber is used without modifying resins. See Table 5-54 for general properties.

Comparison of Cyclized Rubber with Nitrocellulose for Paper Coatings

Advantages	Disadvantages
Much lower density	More than twice as expensive on a volume basis
Soluble in cheap hydrocarbon solvents	Oxidizes and tends to become more brittle with age
Better heat-seal properties	Dries more slowly
Noninflammable	More tendency to block
Better moisture-vapor resistance	Poorer mechanical strength and flexibility
Much better resistance to acids, alkalis, and alcohol	Much poorer resistance to oils
Requires little plasticizer	
Simplifies solvent recovery, because it is generally used in a single solvent	

Comparison of Cyclized Rubber with Vinyl Chloride—Vinyl Acetate Copolymer for Paper Coatings

Advantages	Disadvantages
Soluble in cheap hydrocarbon solvents	About twice as expensive on a volume basis
Much better moisture-vapor resistance	Oxidizes and tends to become more brittle with age
Easier to handle in solution, because it has no thixotropic or gel behavior	Much poorer mechanical strength and flexibility
Does not contain chlorine which can split off as HCl	Much poorer resistance to oils
Requires little plasticizer	

Zein

Chemical Constitution. Zein is a complex protein polymer produced from corn by extracting gluten meal with isopropyl alcohol.[16] Its molecular weight is reported to be approximately 40,000.

Solubility. The most commonly used solvents are 95% ethyl alcohol and 91% isopropyl alcohol. Anhydrous alcohols, above methanol, do not dissolve zein. Anhydrous methanol is a solvent, but gives solutions of poor stability. Other useful solvents are Cellosolve and methyl Cellosolve. Acetone, methyl ethyl ketone, Cellosolve acetate, nitromethane, and nitropropane are latent solvents. Aromatic and aliphatic hydrocarbons can be used in small proportions as diluents.

Plasticizer Compatibility. The most commonly used plasticizers are

are stearic, oleic, and palmitic acids; the glycols; and ethyl toluene sulfonamide. Dibutyl phthalate, dioctyl phthalate, and tricresyl phosphate are not good plasticizers by themselves, but are useful when used together with the plasticizers previously listed.

Resin Compatibility. Zein is compatible with rosin, hydrogenated rosin, terpene resins, manila gum, urea–formaldehyde resins, alcohol-soluble phenolics, and cumar resins.

General Remarks. The most common application of zein solutions in the paper field is for greaseproof and heat-sealing coatings.

Secondary Film Formers–Resins

Resins are added to primary film-forming materials for several reasons:

> To increase the solids concentration without commensurate increase in viscosity
> To improve the gloss
> To improve adhesion
> To improve special properties, such as heat sealing, grease resistance, and chemical resistance.

The incorporation of some resins into the coating may also be accompanied by undesirable effects, which may offset some of the advantages previously outlined. Hard, brittle resins decrease the flexibility of the coating and markedly affect the folding properties, especially on a filled or precoated sheet. If enough plasticizer is added to compensate for this increased brittleness, the coating may have poor blocking properties. However, if softer, low-melting resins are used, the final flexibility is satisfactory, but solvent retention, blocking tendency, and viscosity are usually increased, although the speed of drying is usually decreased.

Soft, plastic resins are often used for their plasticizing properties, and sometimes produce films of better mechanical properties than those containing chemical plasticizers. A resin used with primary film formers for paper coatings should have as many of the following properties as possible:

> Good compatibility with the other ingredients
> Good solubility in the solvent combination necessary for the primary film former
> Insolubility in water
> Freedom from toxicity
> Reasonable cost

TABLE 5-55

Representative Resins and Their Compatibilities with Primary Film Forming Materials

Resin	Source	Type	Melting point, °C	Compatibility
Acryloid C-10	Resinous Products	Acrylic Ester	Solution	1, 6
Aroplaz 930	USI		Solution	1, 5, 10
Beckosol 1323	Reichhold		Solution	1, 5
Paraplex RG-2	Resinous Products	Alkyd–Nonoxidizing	Solution	1, 5, 12
Paraplex G-20	Resinous Products		Solution	1, 4, 7
Rezyl X315	Am. Cyanamid		Solution	5, 10
Aroplaz 905	USI		Solution	1, 5
Beckosol P-94	Reichhold		Liquid	5
Duraplex A-25	Resinous Products	Alkyd–Oxidizing	Solution	5, 7
Esterol 575	Paramet Chemical		Solution	5
Rezyl 807	Am. Cyanamid		Solution	5, 10
Amberlac D-96	Resinous Products		98—100	1, 5, 7
Amberol 801	Resinous Products		142—149	1, 5, 10
Beckacite 1110	Reichhold	Alkyd–Modified	108—113	5
Lewisol 18	J. D. Lewis Inc.		95—98	5
Paranol Ib-89	Paramet Chemical		82—89	5
Teglac 15	Am. Cyanamid		99—101	1, 5, 6, 10, 4
Nevillite No. 1	Neville Co.	Cyclo Paraffin	100—107	5, 10, 11
Clorafin 42	Hercules Powder	Chlorinated Paraffin	Liquid	9, 10
Cumar P-10	Barrett	Coumarone–Indene	Plastic	5, 7, 10, 11
Neville R-9	Neville Co.	Coumarone–Indene	101	5, 10
		Phenol-Formaldehyde		
Amberol F-7	Resinous Products	Modified Phenolic Phenol-Formaldehyde	148—156	5

Product	Manufacturer	Type	Melting Range	Compatibility
Beckacite 1100	Reichhold	Modified Phenolic	136-144	5
Bakelite BR-254	Bakelite Corp.	Phenol-Formaldehyde	82-88	5, 6, 7, 3, 4
Super Beckacite 1001	Reichhold	Pure Phenolic	72-77	2, 5, 6, 10, 7
Super Beckacite 2000	Reichhold		81-83	2, 5, 9
Santolite MS	Monsanto	Toluene-Sulfonamide	Soft	1, 2, 5, 6, 7, 9, 12
Beckamine P-138	Reichhold	Urea Formaldehyde	Solution	1, 5, 10
Uformite F-220-E	Resinous Products	Urea Formaldehyde	Solution	1, 5
Accroides	Am. Gum Importers	Natural Gum	124-1335	5, 6, 7
Batavia Dammar	—	Natural Resin	85-93	3, 5, 10, 11
Manila Copals		Natural Resin	96-104	5, 6, 10
Resin WW	Hercules Powder	Natural Resin	62-68	5, 6, 7, 10
Shellac S-99	A. Hurst & Co.	Natural Resin	70-78	5, 6, 12
Vinsol	Hercules Powder	Processed Natural Resin	82-89	2, 3, 5, 6, 10, 4
Abalyn	Hercules Powder	Rosin Ester	Liquid	5, 9, 3 4, 11
Ester Gum 7	Am. Cyanamid	Rosin Ester	86-92	2, 3, 7, 9, 10, 11
Aroclor 1254	Monsanto	Chlorinated Diphenyl		1, 3, 6, 7, 9, 10, 4

Code for Compatibility

1-Nitrocellulose
2-Cellulose acetate
3-Cellulose aceto-butyrate
4-Cellulose aceto-proprionate
5-Ethyl cellulose
6-Vinyl acetate
7-Vinyl chloride–vinyl acetate copolymer
8-Vinylidene chloride–vinyl chloride copolymer
9-Vinylidene chloride–acrylonitrile copolymer
10-Chlorinated rubber
11-Cyclized rubber
12-Polyvinyl butyral

Stability against decomposition under conditions of use
Light color
No undesirable reactivity
Minimum solvent retention
Minimum tendency to increase brittleness

Both synthetic and natural resins are used in paper coatings. The principal types used for this purpose are discussed in the following.

Nonoxidizing, Oxidizing, and Modified Alkyds

Alkyd resins are made by the reaction of polybasic acids or anhydrides, such as phthalic, maleic, succinic, or sebasic, with polyhydric alcohols, such as glycerine or glycols. They are often modified with fatty or resin acids and saturated and unsaturated oils. They may be modified also with phenol and urea–formaldehyde. Depending on the modification and the polybasic acids and polyhydric alcohols used, the resulting resins may be thermoplastic, thermosetting, oxidizing, or nonoxidizing.

Pure and Modified Phenolic Resins

These resins are made principally by the condensation of phenol and formaldehyde. A great many resins are possible. The phenol can be replaced by substituted phenols, such as p-phenylphenol, p-tertiary butylphenol, or isobutylphenol. The formaldehyde can be replaced wholly or in part by other aldehydes, such as acetaldehyde and butyraldehyde. Rosin, ester gum, polyhydric alcohols, fatty acids, and drying oils can also be added, and enter into chemical reaction with, or are merely dispersed in, the resin.

Aldehyde–Amine Resins

The most important resin of the aldehyde–amine class is the urea-formaldehyde type. Other amines used are thiourea, guanidine, aromatic amines, such as aniline and substituted anilines, and toluene sulfonamide. Other aldehydes besides formaldehyde can also be used in the resin reaction.

Acrylic Resins

These are polymers or copolymers of acrylic and methacrylic acid derivatives, chiefly of the methyl and ethyl esters of these acids. Other esters of these acids, such as the propyl and butyl, can also be used.

Coumarone–Indene Resins

These are made by polymerizing the coumarone and indene obtained from coal-tar naphthas produced in by-product coke ovens. Several grades are made, varying from very soft to hard resins.

Natural Resins

These include accroides, elemi, sandarac, dammar, East India, copal, rosin, and shellac.

Rosin Esters and Hydrogenated-Rosin Esters

The most important type of rosin ester is ester gum—the glycerol ester of rosin. Other modified rosin types are the methyl ester, hydrogenated methyl ester, diethylene glycol ester, glycerol and polyglycol esters of hydrogenated rosin.

Vinyl Resins

These have been included arbitrarily under Primary Film Formers. They may, however, be used also as modifying resins for other primary film formers with which they are compatible.

Table 5-55 contains a list of representative resins, showing trade names, manufacturers, types, melting points, and compatibilities with primary film formers. Lack of compatibility data does not necessarily mean that the resin is incompatible, but only that the data were not available. Many additional resins commercially available, may be equally suitable, but have not been listed here because of limitations of space.

Plasticizers

Plasticizers are added to the nonvolatile portion of the film to improve its flexibility. They also reduce the shrinkage of the coating and lessen the tendency of the coated paper to curl. For every primary film former, a definite optimum quantitative range of plasticizer can be used in coatings for paper. The proportion that should be used depends on the primary film former, the type of plasticizer, the amount and type of added resin, and the degree of pigmentation. As the proportion of resin is increased, extra plasticizer must be added over and above the amount required for the primary film former alone. As the pigmentation is increased, the plasticizer must also be slightly increased. As the proportion of plasticizer in a given formulation is increased, the tensile strength of the film and the degree of curl decrease, whereas the elongation of the film increases and the folding properties improve. A point is reached, however, where the film becomes excessively soft, with accompanying increasing degrees of tackiness. This tackiness at first may be only very slight, as evidenced by a certain lack of slip when the coated surfaces are rubbed together. On further addition of plasticizer, this lack of slip or slight tackiness increases, and the coating becomes more and more sticky.

The correct proportion of plasticizer corresponds to the amount that produces the best flexibility and noncurl without producing tackiness in the film. This proportion can be found very readily by coating strips of thin paper (0.003—0.004 in.) with a series of coatings in which the plasticizer content is progressively increased. With the correct amount of plasticizer, the sheet will just lie flat without curling when thoroughly dry. In a given formulation, this range is very narrow, and the test will often give the best proportion of plasticizer within 5%, based on the amount of primary film former used. The determination of this narrow range may not always completely solve the plasticizing problem. It may well happen, with certain types of paper and with certain primary film formers, that at the point where no more plasticizer can be added for fear of blocking the folding properties are still not as good as desired and there is no freedom from curl. This point, however, is the best possible compromise under these conditions.

Plasticizers are mobile liquids, oils, soft plastic resins, and sometimes solids. There are two general classes: solvent type and nonsolvent type. In some cases, either type can be used. In other cases, solvent types are the most suitable and in still other cases, a combination of the two types is preferable. The nonsolvent types are probably dispersed in the film in minute globules and can be forced out to some extent by heat and pressure. The solvent-type plasticizers usually have a greater effect in lowering the tensile strength and increasing the elongation. Every plasticizer has its own characteristic stress–strain curve when used with a particular primary film former. Nonsolvent plasticizers have less tendency to produce tackiness and thermoplasticity, whereas solvent plasticizers have less tendency to separate out of the film and usually are more stable chemically. The solvent plasticizers are generally chemicals, while the nonsolvent plasticizers are oils. However, exceptions are common. Castor oil, for example, is a nonsolvent plasticizer for nitrocellulose, but a solvent plasticizer for ethyl cellulose.

An ideal plasticizer should have the following properties:
    High plasticizing efficiency over a wide range
    Compatibility with the primary film former and resin used
    Good solubility in the combination of solvents used
    Low volatility

Insolubility in water
Freedom from odor
Water-white color
Chemical stability under conditions of use
Nontoxicity for specialized applications
No tendency to yellow
Good resistance to sunlight
Low acidity
Low inflammability

A characteristic of plasticizers, often encountered under certain conditions, is migration, i.e., a tendency to diffuse through the film deposited by multiple coats, where varying proportions of plasticizers, their combinations, or dissimilar plasticizers, have been used in the several coats. This condition is often accelerated by heat and pressure. If a dry, moderately plasticized film is deposited over a highly plasticized film, there is often a tendency for the plasticizer in the bottom coat to migrate to the top and produce a tacky surface. Both solvent and nonsolvent plasticizers are subject to this action. Migration often occurs when two dissimilar films are superimposed. For example, if a plasticized nitrocellulose film is deposited on top of a plasticized cellulose acetate film, the surface may eventually become tacky, even though both films were perfectly dry originally. Migration may also occur when two objects finished with two dissimilar coatings are left in contact with each other for some time. When a paper is saturated with a plasticized compound and surface coated with a plasticized solvent coating, migration from the interior of the paper to the coated surface may take place.

TABLE 5-56

Plasticizer Requirements of Various Film Forming Resins

Primary film former	Percent by weight based on primary film former
Nitrocellulose	65-100
Cellulose Acetate	60-80
Cellulose Acetobutyrate	60-70
Cellulose Acetopropionate	60-70
Ethyl Cellulose	25-35
Polyvinyl Acetate	10-15
Vinyl Chloride–Vinyl Acetate Copolymer	15-25
Vinylidene Chloride–Acrylonitrile Copolymer	0-15
Vinyl Chloride–Vinylidene Chloride Copolymer	15-25
Polyvinyl Butyral	20-40
Chlorinated Rubber	20-25
Cyclized Rubber	0-10

TABLE 5-57

Plasticizers

Name	Trade name	Molecular weight	Boiling range °C	Compatibility
Methyl Abietate	Abalyn	316	360-365 Decomposes	1, 5, 7, 12
Hydrogenated Methyl Abietate	Hercolyn	318	365-370 Decomposes	15, 7, 10, 12
Chlorinated Diphenyl	Aroclor 1242	—	322-365	1, 3, 5, 6, 7, 10, 11, 12
Camphor		152	204	1, 10
Triethyl Citrate		276	150 at 3 mm	1, 2, 3, 4, 5, 6, 7, 10, 12
Glycerol Triacetate	Triacetin	218	259-262	1, 2, 3, 5, 6
Triethylene Glycol Di-2-Ethyl Butyrate	Flexol 3GH	346	202 at 5 mm	1, 5, 6, 7
Methyl Phthalyl Ethyl Glycolate	Santicizer M-17	266	189 at 5 mm	1, 2, 3, 4, 5, 6, 7, 10, 12
Butyl Phthalyl Butyl Glycolate	Santicizer B-16	336	219 at 5 mm	1, 3, 5, 6, 7, 8, 9, 10, 12
Diethylene Glycol Monolaurate	Glaurin	288		1, 5, 7, 12
Amyl Oleate		353	200-240 at 20 mm	1, 5, 10, 12
Tetrahydrofurfuryl Oleate		366	200-285 at 16 mm	1, 5, 6, 7
Paraffin Wax				11
Chlorinated Paraffin	Clorafin 42	593		5, 6, 7
Tributyl Phosphate		266	177-178 at 27 mm	1, 3, 5, 7, 10, 12
Triphenyl Phosphate		326	220 at 5 mm	1, 2, 3, 4, 5, 6, 9, 10
Tricresyl Phosphate	Lindol	368	235-255 at 4 mm	1, 3, 5, 6, 7, 9, 10, 11, 12
Dimethyl Phthalate		194	282	1, 2, 3, 4, 5, 6, 7, 10, 12
Diethyl Phthalate		222	295	1, 2, 3, 4, 5, 6, 7, 9, 10, 12
Dibutyl Phthalate		278	339	1, 3, 5, 6, 7, 9, 10, 12
Diamyl Phthalate		306	342	1, 3, 5, 6, 7, 10, 12
Dicapryl Phthalate	Monoplex DCP	390	227-234 at 4.5 mm	1, 3, 5, 7, 8, 12
Diethyl Hexyl Phthalate	Flexol DOP	390	386	1, 5, 7, 8, 12
Dimethoxyethyl Phthalate	Methox	282	190-210 at 4 mm	1, 2, 3, 4, 5, 6, 7, 12

				Code for Compatibility
Dibutoxy Ethyl Phthalate	Kronisol	366	210-235 at 4 mm	1, 3, 5, 6, 7, 8, 12
Esterified Coconut Oil	Plasticizer SC	399	210-226 at 7 mm	7, 8, 12
Castor Oil				1, 10, 11, 12
Methyl Ricinoleate	Baker P-1	312	225 at 15 mm	1, 3, 5, 6, 12
Methyl Acetyl Ricinoleate	Baker P-4	354		1, 5, 6, 7, 12
Dibutyl Sebacate	Monoplex DBS	314	344-345	1, 3, 5, 6, 7, 10, 12
Dibenzyl Sebacate	Monoplex 5	382	265 at 4 mm	1, 5, 6, 9, 12
Butyl Stearate		340	220-225 at 25 mm	1, 5, 10, 11
1, 1-Diethyl Stearamide		339	190-205 at 1 mm	1, 5, 7, 12
Dibutyl Succinate		230	255	1, 2
o-Cresyl p-Toluene Sulfonate	Santicizer 10	262		1, 3, 5, 6
N-Isopropyl Benzene Sulfonamide	Santicizer 130	199		1, 2, 3, 4, 5, 6
O-and-P-Toluene Ethyl Sulfonamide	Santicizer 8	199		1, 2, 3, 4, 5, 6, 12
Dibutyl Tartrate		262	292-312	1, 2, 3, 4, 5, 6, 7
Tung Oil				10, 11
Diphenyl Mono-oXenyl Phosphate	Dow # 5	402	250-285 at 5 mm	1, 3, 5, 6, 7, 12

Code for Compatibility

1-Nitrocellulose
2-Cellulose acetate
3-Cellulose aceto-butyrate
4-Cellulose aceto-proprionate
5-Ethyl cellulose
6-Vinyl acetate
7-Vinyl chloride–vinyl acetate copolymer
8-Vinyl chloride–vinylidene chloride copolymer
9-Vinyl chloride–acrylonitrile copolymer
10-Chlorinated rubber
11-Cyclized rubber
12-Polyvinyl butral

Little information, either qualitative or quantitative, is available on this problem. Every application where multiple coats with varying plasticizer content or with varying primary film formers are used should be carefully checked by an accelerated test for migration. A useful accelerated test involves the exposure of the coating to a temperature of 75—100°C, preferably at high humidity, for a period of several days. This test is not infallible, however, and if a plasticizer of relatively high volatility is present, much of it will evaporate under these conditions and give erroneous results.

Table 5-56 lists the approximate percentages of plasticizer that produce optimum results in clear coatings on a light-weight sheet of clay-coated and supercalendered paper. This percentage is subject to some variation, depending on the type of plasticizer used and on the viscosity of the primary film former, and must be modified somewhat when resins and pigments are added.

Table 5-57 lists a number of the most common plasticizers, giving their usefulness and compatibility with the primary film formers. This table includes plasticizers mentioned already in the discussion of the various primary film formers. For more detailed data and further tables on plasticizers listing their physical and chemical properties, reference should be made to the literature.[5,17,18,19,20,21]

Pigments, Dyes, Fillers, and Bronze Powders

Pigments and dyes are used in the coating to produce the desired color. Fillers are used to extend the pigment, increase the solids, reduce tackiness, or reduce the gloss. The addition of pigments and fillers to the common primary film formers usually decreases both the tensile strength and the elongation, and also the adhesion to the paper surface. It also decreases the gloss. Some of the desirable properties of a pigment are:

High opacity and hiding power
Good fastness to light
Good stability
Nonreactivity
Insolubility in water
Ease of grinding or dispersion
Reasonable cost
Minimum tendency to seed or settle
Freedom from bleeding or bronzing

Most pigments do not possess all these properties, and the best possible compromise has to be made for every application.

The proportion of pigment used in the coating will vary greatly, depending on the type of pigment, the primary film former, and the gloss desired. The quantitative range is from 5—100%, by weight, based on the primary film former. Approximately 25% is the maximum proportion for a high-gloss coating. For matte coatings, 50—100% pigment is required for good flatness if no special flatting agents are used. With low-viscosity polymers, the proportion of pigment that can be added without sacrificing good flexibility is much less than with high-viscosity polymers. For example, 50% titanium dioxide with ½-sec nitrocellulose, properly plasticized, produces a film of relatively poor folding properties. When properly plasticized 30—40-sec nitrocellulose is pigmented to the same extent and the folding is much better.

The pigment or filler must be properly ground or dispersed before

TABLE 5-58

Commonly Used Pigments, Fillers, and Flatting Agents

Whites
Titanium Dioxide
Titanium Barium Pigment
Titanium Calcium Pigment
Lithopone
Zinc Oxide

Yellows and Oranges
Chrome Yellow and Chrome Orange
Zinc Yellow
Cadmium Yellow
Hansa Yellow
Molybdenum Orange
Yellow Iron Oxides

Reds and Maroons
Iron Oxide, and Spanish Oxide
Tuscan Red
Lithol Red
Toluidine Red
Para Red
Cadmium Red
Scarlet Lake
Pigment Scarlet Lake
Lithol Rubine Toner

Blues
Ultramarine Blue
Iron Blue
Copper Phthalocyanine Blue

Blacks
Carbon Black
Bone Black
Lamp Black

Greens
Chrome Green
Chromium Oxide Greens
Green Toners and Lakes

Browns
Brown Oxide
Siennas
Umbers

Flatting Agents
Aluminum Stearate
Calcium Stearate
Zinc Stearate
Anhydrous Silica

Fillers
Barytes
Blanc Fix
Clay
Calcium Carbonate
Silica
Magnesium Carbonate
Talc

it is incorporated into the coating solution. It may be ground either in a portion of the plasticizer or in a portion of the vehicle. Poorly dispersed pigment particles show up conspicuously in the finished coating. They not only contribute to the formation of scratches in the process of coating, but they detract from the appearance and interfere with any printing on the sheet. They also act as focal points for starting breaks in the film when the paper is folded.

Dyes are often used to obtain brilliant shades that are not obtainable with pigments. These dyes must be solvent soluble. They are first dissolved, in a concentration of 4—8 oz/gal of solvent, and filtered through paper to remove insoluble matter. The proportion of dye used depends on the shade, and it may vary from less than 1% up to 5% by weight of dry dye, based on the primary film former. These dyes are seldom fast to light, and have low hiding power. For this reason, they are best used together with pigment colors.

Flatting agents and fillers contribute little or no opacity to the coating. The former are used to reduce gloss and to produce matte finishes.

Bronze powders are made from aluminum, copper, and copper-zinc alloys. They are incorporated into coatings to produce metallic effects. In aluminum coatings, 50—75% powder by weight, based on the primary film former, is ordinarily used. In copper and bronze coatings, the proportion is usually 100—150%.

For a detailed discussion and tabulation of pigments, dyes, fillers, and bronzes, the reader is referred to the literature.[22,23]

SOLVENTS, LATENT SOLVENTS, AND DILUENTS

A volatile organic liquid may be classified as a solvent, latent solvent, or diluent, with respect to its solvency action on a particular film-forming compound. It is called a solvent if it dissolves the film-forming compound by itself. Before the terms *latent solvent* and *diluent* are defined, the meaning of the term solvency must be clarified. Solvency is measured by a combination of two properties—solution viscosity and tolerance. Viscosity is a familiar concept; it is a measure of the internal resistance of a liquid or a solution to flow. The viscosity of a solution of a given film former in a particular solvent increases with increasing concentration. The viscosity–concentration curve is seldom linear, except at low concentrations. An increase in the concentration usually results in a more-than-proportional increase in viscosity. In solutions of the same film former—at the same concentrations in a homologous series of solvents—the viscosity usually increases with increasing molecular weight of the solvent. For example, a 15-g/100 cm³ solution of nitrocellulose in butyl acetate has a higher viscosity than a solution of the same concentration in ethyl acetate. Similarly, a 15-g/ 100 cm³ solution of nitrocellulose in amyl acetate has a higher viscosity than a solution of the same concentration in butyl acetate. From the standpoint of viscosity, ethyl acetate can be said to have better solvency for nitrocellulose than butyl acetate or amyl acetate.

If a portion of the solvent is replaced by an equal volume of non-solvent, the solution viscosity may remain the same, decrease, or increase. The effect on viscosity of this substitution depends on the film former, its concentration, the type of solvent and nonsolvent used, and the degree of replacement. In some cases, even the slightest substitution results in a higher viscosity, which continues to increase with increasing degrees of substitution of nonsolvent for solvent. In other cases, the viscosity drops at first and continues to decrease until a definite degree of substitution is reached. It then rises rapidly to the point of maximum substitution. Bogin has plotted a series of curves showing the viscosity relationships of nitrocellulose solutions in mixtures of common lacquer solvents with toluol and petroleum naphtha.[24] Two curves are of particular interest for this discussion. One curve is a plot of the viscosities of nitrocellulose in mixtures of Cellosolve and toluol;the other is a plot of the viscosities of nitrocellulose in mixtures of ethyl acetate and petroleum naphtha. Both curves show minimum turning points. In the first, a minimum viscosity is reached when 33% of the solvent has been replaced by toluol. In the second, a minimum viscosity is reached when 20% of the solvent has been replaced by naphtha.

As the degree of substitution of solvent by nonsolvent increases, a point is reached where the film former begins to precipitate. The term *tolerance* refers to the volume ratio of nonsolvent to solvent at this point. Tolerance depends not only on the nature of the solvent, nonsolvent, and film former, but also on the end concentration, temperature, and the presence of resins and plasticizers. Tolerance values must be compared at the same end-point concentration and temperature to have any real meaning. Solvents can be graded as to solvency from tolerance considerations.

As for the question of defining the terms latent solvents and diluents, these definitions could be made on the basis of the effect of a liquid on either viscosity or tolerance. A liquid could arbitrarily be called a latent solvent if its partial substitution for a true solvent results in no

TABLE 5-59

Properties of Common Lacquer Solvents

Solvent	Chemical formula	Molecular weight	Boiling point of pure material, °C	Vapor pressure at 20°C, mm Hg	Heat of vaporization, g cal at 30°C	Evaporation rate by weight (n-butyl acetate=100)	lb/gal at 20°C	Flash point (Cleveland open cup), °C	Maximum allowable conc. for constant exposure, ppm	Explosive limits, % by vol. in air Upper	Explosive limits, % by vol. in air Lower
Methyl Alcohol	CH_3OH	32	64.5	98.5	276.0	370	6.6	15	200[1]	36.5[4]	6.0[4]
Ethyl Alcohol	CH_3CH_2OH	46	78.5	44.0	216.2	203	6.6	21	1000[2]	19[4]	3.28[4]
n-Propyl Alcohol	$CH_3CH_2CH_2OH$	60	97.2	14.5	164 at 97°C	205	6.7	22	400[2]	2.5[4]	2.5[4]
Isopropyl Alcohol	$(CH_3)_2CHOH$	60	82.5	33.0	184.0	45	6.58	21			2.5[4]
n-Butyl Alcohol	$CH_3CH_2CH_2CH_2OH$	74	117.1	5.5	161.0	83	6.75	46	50[1]	18[5]	1.7[4]
Isobutyl Alcohol	$(CH_3)_2CHCH_2OH$	74	107.9	99 at 60°	147.0	115	6.68	27			1.68[4]
Sec-Butyl Alcohol	$CH_3CH_2CHOHCH_3$	74	99.5	12.1			6.72	24			
n-Amyl Alcohol	$CH_3CH_2CH_2CH_2CH_2OH$	88	138	2.8	134 at 98°C		6.83	57			1.2[4]
n-Hexyl Alcohol	$CH_3CH_2CH_2CH_2CH_2CH_2OH$	102	157.2	0.7	121.2 at 131°C		6.93	74			
2-Ethylbutyl Alcohol	$(C_2H_5)_2CHCH_2OH$	102	148.9	0.9			6.93	57			
2-Ethylhexyl Alcohol	$CH_3CH_2CH_2CH_2CH(C_2H_5)CH_2OH$	130	183.5	0.36			6.94	85			
Methyl Acetate	CH_3COOCH_3	74	59	169.8	114 at 0°C	1040	7.78	−7	100[2]	13.9[4]	4.1[4]
Ethyl Acetate	$CH_3COOC_2H_5$	88	77.1	65	97.0	525	7.37	4	400[1]	11.5[4]	2.18[4]
Isopropyl Acetate	$CH_3COOCH(CH_3)_2$	102	89.0	47.5	87.0	435	7.28	16			2.0[4]
n-Butyl Acetate	$CH_3COOC_4H_9$	116	126.5	8.7	83.0	100	7.29	38	200[1]	15[4]	1.7[4]
Isobutyl Acetate	$CH_3COOCH(CH_3)_2$	116	118.3	16.4		152	7.16	25			
Sec-Butyl Acetate	$CH_3COOCH(CH_3)CH_2CH_3$	116	111.5	16.2		180	7.16	19			
n-Amyl Acetate	$CH_3COOC_5H_{11}$	130	147.6				7.17	26	200[1]		
Methyl Cellosolve Acetate	$CH_3COOCH_2CH_2OCH_3$	118	143.0	3.3	91.5	40	8.37	60	100[2]		1.1[4]
Cellosolve Acetate	$CH_3COOCH_2CH_2OC_2H_5$	132	156.2	1.2		24	8.10	66	100[1]	15.25[6]	1.71[4]
Ethyl Lactate	$CH_3CHOHCOOC_2H_5$	118	155.0	2.7		22	8.60	61			
Butyl Lactate	$CH_3CHOHCOOC_4H_9$	146	188.0	0.4		6	8.61	81			
Dioxane	$OCH_2CH_2OCH_2CH_2$	88	101.1	29		55	8.61	18	500[7]	22.25[7]	1.97[7]
Methyl Cellosolve	$CH_3OCH_2CH_2OH$	76	124.5	6.2	127	40	8.04	46	100[2]		2.6[4]
Cellosolve	$C_2H_5OCH_2CH_2OH$	90	135	3.8	124.0	10	7.75	57	200[2]	15.7[4]	
Butyl Cellosolve	$C_4H_9OCH_2CH_2OH$	118	170.6	0.6	96.5		7.50	74			
Ethylene Dichloride	$ClCH_2CH_2Cl$	99	83.5	62			10.45	15	75[1]	15.9[4]	6.2
Propylene Dichloride	$CH_3CHClCH_2Cl$	113	96.3	40			9.70	17		14.5[4]	3.4[4]

1,1,2-Trichloroethane	$CHCl_2CH_2Cl$	133.5	113.7	16.7			12.0	None			
n-Pentane	$CH_3CH_2CH_2CH_2CH_3$	72	36.0	425			5.23	<−45	5000[2]	8.0[4]	1.44
n-Hexane	$CH_3CH_2CH_2CH_2CH_2CH_3$	86	68.7	155			5.65	<−25		6.90[4]	1.25[4]
n-Heptane	$CH_3CH_2CH_2CH_2CH_2CH_2CH_3$	100	98.4	35.5	89.1 at 0°C		5.71	<0		6[4]	1[4]
Isooctane	C_8H_{18}	114	102—113	35			6.23	<−5			
Benzene	$CH=CHCH=CHCH=CH$	78	79.6	118 at 30°C	102.2	500	7.51	−8	35[1]	8[4]	1.44
Toluene	$C_6H_5CH_3$	92	111.0	36.7 at 30°C	98.0	195	7.22	6.5	200[1]	7.0[4]	1.27[4]
Xylene	$C_6H_4(CH_3)_2$	106	135—145	11 at 30°C		68	7.2	29	200[1]	5.3[5]	1.04 (o-xylene)
Acetone	CH_3COCH_3	58	56.1	186	130.0	720	6.6	−9	500[1]	13.0[4]	2.15[4]
Methyl Ethyl Ketone	$CH_3COCH_2CH_3$	72	79.6	72	112.0	465	6.72	2	300[1]	11.5[4]	1.81[4]
Methyl Isobutyl Ketone	$CH_3COCH_2CH(CH_3)_2$	100	118.0	15.7	93.2	145	6.67	24			
Methyl n-Amyl Ketone	$CH_3COC_5H_{11}$	114	150.6	2.6	82.7 at 149°C		6.80	49			
Mesityl Oxide	$(CH_3)_2C=CHCOCH_3$	98	128.7	8.7	152.0	87	7.12	32	50[2]		
Cyclohexanone	$CH_2CH_2CH_2CH_2CH_2CO$	98	155.6	10 at 35°C		25	7.87	50	100[3]		
Isophorone	$COCH=C(CH_3)CH_2C(CH_3)_2CH_2$	138	215.2	0.3		15	7.68	96	25[2]		
Diacetone Alcohol	$(CH_3)_2COHCH_2COCH_3$	116	166.0	0.84	55.0	180	7.79	68	200[3]		
Nitromethane	CH_3NO_2	61	101	27.8	135 at 100°C	145	9.5	44			
Nitroethane	$C_2H_5NO_2$	75	114	15.6			8.8	41			
1-Nitropropane	$CH_3CH_2CH_2NO_2$	89	132	7.5		100	8.4	49			
2-Nitropropane	$CH_3CHNO_2CH_3$	89	120	12.9		124	8.3	39.5			

1. H. B. Elkins, "Toxic Fumes," Mass. Dept. of Occupational Hygiene.
2. W. A. Cook, "Maximum Allowable Concentrations of Industrial Atmospheric Contaminants," Ind. Medicine (1945).
3. C. D. Bogin, Protective and Decorative Coatings, Matiello, Vol. 1, p. 644.
4. "Properties of Flammable Liquids, Gases, and Solids," Associated Factory Mutual Fire Insurance Companies, Boston, Mass. (1945).
5. "Fire Hazard Properties of Certain Flammable Liquids," Natural Fire Protection Association, Boston, Mass. (1934).
6. UCC Detector Test Data.
7. Ind. Eng. Chem., 25, 1285 (1933).

increase in solution viscosity. It could arbitrarily be called a latent solvent if some of its mixtures with a true solvent have a similar tolerance for other nonsolvents as the true solvent alone. Common usage favors the second definition for classifying liquids as latent solvents or diluents. A liquid, therefore, is commonly called a latent solvent for a particular film former if it produces no appreciable decrease in the tolerance of the mixture into which it is introduced in certain limiting proportions. Sometimes the tolerance is even increased at certain degrees of substitution. A liquid is a diluent if its mixtures with a true solvent always have lower tolerance values than the solvent alone. These classifications are valid only with reference to the action of the liquid on a particular film former. A liquid that is a diluent for one material may be a solvent for another. For example, toluol is a solvent for polyvinyl acetate, but a diluent for nitrocellulose. In many cases, a mixture of alcohol and ether is a solvent for nitrocellulose, although neither liquid alone dissolves it. To cover this type of case, the definition of a latent solvent may be broadened to include any compound that by itself does not dissolve a given film-forming material, but that, in the presence of some other substance, exhibits the properties of a true solvent.

Organic solvents used for paper coatings fall into the following chemical classes:

Esters
Alcohols
Ketones
Aliphatic and Aromatic Hydrocarbons
Chlorinated Hydrocarbons
Nitroparaffins
Glycol Ethers
Ethers and Oxides

By far the largest number of solvents used in the paper-coating field is in the first four groups.

The usefulness of an organic liquid as a solvent depends on some of the following characteristics:

Solvency properties for primary and secondary film formers
Viscosity vs solid concentration
Effect on thixotropic behavior and gel formation
Evaporation rate
Effect on blushing and flow
Odor—both initial and residual
Toxicity

Flash point and explosive limits
Cost
Ease of recovery

Effect of Solvent on Film Properties

The primary function of the solvent mixture is to convert the non-volatile components into a physical state of mobility so that they can be applied as a coating to the desired surface. Although the solvent leaves the deposited wet film entirely, by evaporating into the air (except small amounts which may be trapped or retained for a considerable time), and does not form part of the resulting dry coating, it does have a decided effect on the film properties. The type of solvent used, its evaporation rate, and its solvency characteristics determine, to a great extent, the smoothness, gloss, adhesion, tensile strength, elongation, and freedom from bubbles and other blemishes of the dry coating. These effects seem entirely logical when it is considered that the solvency characteristics determine the state in which the soluble compounds are dispersed and that the rate of evaporation determines the rate of viscosity increase in the drying film and, consequently, the rate at which the nonvolatile components coalesce to form a continuous film. This is not merely a theory, but it is substantiated in everyday practice. In the case of solution coatings made with ethyl cellulose, for example, the the differences in tensile strength and elongation of films deposited from different solvent combinations are striking.

Solvency and Its Relation to Thixotropic and Gel Phenomena

With film formers such as nitrocellulose and other cellulose derivatives, viscosity and tolerance are the only solution characteristics of great importance. With vinyl resins, however, thixotropic and gel behavior must also be considered. With a given solvent and diluent, a curve can be plotted of the maximum concentration of a vinyl resin vs the percentage of diluent at any given temperature, before pronounced thixotropic behavior becomes evident. A similar curve can be drawn with respect to gel formation. Such curves for solutions of vinyl chloride–vinyl acetate copolymer in mixtures of methyl ethyl ketone and toluol have been developed. In formulating solvent mixtures for the copolymer these curves are very important guides. In the booklet, *Vinylite Copolymer Resins for Surface Coatings*, published by the Bakelite Corporation, a series of such phase diagrams for various solvent mixtures is given.[7]

Evaporation Rates of Solvents

One of the most important properties of a volatile liquid used in a solvent mixture is its rate of evaporation. Volatile liquids used in the lacquer industry have often been erroneously classified as "low boiling," "medium boiling," and "high boiling." The intention of this classification is really to convey an idea of the rate of evaporation or drying. In most cases, rates of evaporation of volatile liquids have no direct relation to their boiling points. Of greater interest is how fast the liquid will evaporate, and not what its boiling point is. The rate of evaporation of a liquid is a function of its *vapor pressure*, its molecular weight, its temperature, and its latent heat. With respect to external conditions, the rate of evaporation of a liquid is a function of the temperature difference between the surroundings and the liquid, the difference vetween the vapor pressure of the liquid and that of the vapor in the adjacent air, and the air velocity passing over the liquid film.

Vapor pressure is measured by the number of moles of vapor that are present in a given volume of air over the liquid surface at equilibrium. If this equilibrium is continually disturbed by air movement, more vapor leaves the surface of the liquid in an attempt to reach equilibrium. The *molal* rate of evaporation is directly proportional to the number of moles that leave the liquid per unit time. The *weight* rate of evaporation, however, is proportional to the product of the molal rate and the molecular weight. In other words, vapor pressures (by themselves) of liquids at a given temperature are not direct measures of evaporation rates. Only when the vapor pressures are multiplied by the molecular weight do they have a direct meaning with regard to weight rate of evaporation. The analysis of evaporation rate is further complicated by a consideration of the latent heat of vaporization. If two hypothetical liquids have identical vapor pressures at a given temperature, and also equal molecular weights but different latent heats, the liquid with the lower latent heat will evaporate faster under the same external conditions. The complication does not stop here, however. When solvents evaporate, the temperature of the liquid drops, since part of the heat of vaporization comes from the liquid itself. As the temperature of the liquid drops, its vapor pressure drops, with a resultant lowering of the weight rate of evaporation.

Simply because one solvent evaporates at a certain relative rate with respect to another solvent, at a given temperature, it will not have the same relative rate at another temperature. Every liquid has its own vapor pressure-temperature curve. The slopes of these curves have their individual characteristics, and in a series of curves for various liquids, some of the curves may intersect.

Laboratory procedure for obtaining evaporation rates is no simple matter and is beset with many pitfalls. It is beyond the scope of this chapter to go into this problem, other than to say that conditions must be very carefully controlled—in particular, the air movement past the evaporationg liquid.

From the drying standpoint, volatile liquids are more significantly classified by evaporation rate than by boiling point. Bogin classifies nitrocellulose solvents as fast evaporating, medium evaporating, slow evaporating, and extra-slow evaporating.[24] According to his classification, fast-evaporating liquids are those with evaporation rates at least three times as great as that of butyl acetate. Medium-evaporating liquids are those with rates greater than one and one-half times that of butyl acetate. Slow-evaporating liquids have rates that are higher than that of Pentosol, but lower than that of secondary butyl acetate. Finally, the extra-slow liquids have evaporation rates lower than that of Pentosol. This classification of course is entirely arbitrary, but is useful for nitrocellulose solvents. Similar classification can be made for volatile liquids applicable with other film formers.

Blushing

A phenomenon often encountered in connection with the use of solvent coatings is "blushing." This term refers to a change of the film from a clear or colored appearance (if it is pigmented) to a white, grayish, or iridescent appearance during drying. This condition may be either temporary or permanent.

Blushing may be of two kinds: water blush and diluent blush. Water blush is caused by the deposition of water in the wet film in the course of drying—for example, as follows.

As the solvent evaporates from the wet film, the temperature of the film drops because of loss of heat. At the same time, heat transfer from the adjacent layer of air to the wet film causes a drop in the temperature of the air. This rate of heat loss is proportional to the difference between the temperature of the air and that of the wet film. A condition may arise where the rate of heat loss of the air is sufficient to reduce its temperature to its dew point. At that point, moisture condenses from the air into the wet film and becomes dispersed in the film in the form of tiny droplets. This condition gives rise to water blush. Why the changes associated with blushing take place under this condition is a subject of some controversy. The general impression is that the white appearance is due to the precipitation of the film former from the solution by the condensed water. Bogin, however, points out that, at least in nitrocellulose solutions, the explanation of precipitation is not valid.[24] He suggests that the white appearance is due to the presence of tiny pores that are at first filled with condensed water, but that become filled with air when the water evaporates. He gives convincing arguments in strong support of this assumption. Whether or not this explanation holds in the case of other film formers is not clear, and should make an interesting subject for study.

Any conditions that promote the condensation of moisture from the air tend to cause this type of blushing. As the dew-point temperature of the air increases with increasing humidity, the greater the relative humidity at a given temperature, the greater tendency there is to produce condensation. Humidity is therefore an important factor in water blush. The evaporation rate of the solvent is another important factor, because it determines the cooling effect on the wet film. The greater the cooling effect, the greater the heat transfer rate from the adjacent air, with a resulting faster temperature drop to the dew point. Another factor involved is the presence of hygroscopic solvents. As they have some tendency to pick up moisture from the air, they tend to promote water blush.

Whether a water blush is temporary or permanent depends considerably on the relation of the evaporation rate of water to that of the slower-drying solvent portion of the wet film. If the water evaporates more slowly than the slowest-drying solvent component, the blush will be permanent. However, if solvents are still present in the film when the water has evaporated, they may bring about sufficient flow to eliminate the condition of blush.

Fortunately, water blush usually is not a serious problem in commercial applications of solvent coatings on paper. The reason for this is that in most cases drying takes place in a chamber at elevated temperatures that do not drop to the dew point, even when rapidly evaporating solvents are used.

A diluent blush, however, is brought about by the appearance of an unfavorable balance in the solvent mixture during the course of evaporation. This unfavorable balance is due to the increase of the proportion of diluent to true solvent above the tolerance point, and its source is a poorly-formulated solvent mixture in which the true solvents leave the film faster than the diluents, with a resulting precipitation of the film former at the critical stage of solvent–diluent balance. Not only the primary film former but also the resin component may precipitate, if a mixture unfavorable for its solution is reached in the course of evaporation.

Toxicity of Solvents and Explosive Hazard

See the section on "Manufacturing Hazards."

Solvent Formulation

In formulating solvent mixtures for paper coatings, solvency, evaporation rate, and blushing tendency must be carefully considered. Other important considerations are toxicity, explosion hazard, and cost, which have not yet been discussed. In general, solvents and nonsolvents of high and medium evaporation rate form the major portion of the volatiles used for paper coatings. In continuous coating, the rate of drying is important and it determines to a large extent the economics of the operation. A fraction of a minute is a very important time interval in the continuous drying of paper. In a 100-ft, single-pass dryer, a coated paper traveling at 100 ft/min must be dry in 1 minute. If the drying requires 2 minutes, the speed must be cut to one-half.

The highest possible evaporation rate consistent with good finished-film properties is the goal to be sought. This is always a compromise between high mechanical properties, smoothness, and high gloss on the one hand, and production speed on the other. Because nearly all paper coating is done by flow methods rather than by spraying, it is possible to take maximum advantage of the use of fast-evaporating solvents for the bulk of the volatile mixture. This is in contrast with spraying operations, where the solution is broken up into droplets, which results in such rapid evaporation that slower-evaporating solvents must be used for good results.

The blending of solvents, latent solvents, and diluents into a solvent mixture usually has for its principal aim the lowest-cost solution with the maximum concentration of nonvolatiles, at the viscosity that is best for the particular method of application and that is consistent with the evaporation-rate and blush-resistance requirements. In a series of solvent mixtures, that with the lowest cost usually is not the one with the maximum concentration at the desired viscosity. Other things being equal, the mixture chosen should be that with the lowest cost of volatiles per unit volume of nonvolatile deposited. In the blending of solvents, maximum advantage should be taken of the incorporation of any lower-cost latent solvents and diluents, if considerations other than cost permit.

Another factor to be kept in mind in solvent formulation is solvent retention in the film. This matter is discussed in the section on drying. It is more prevalent in films containing large amounts of resins and vinyl derivatives. Although slow-evaporating solvents of course tend to stay in the film longest, it must be noted that rapid evaporation may cause skin drying, which will trap solvents in the film. It may be advisable, under certain conditions, to decrease the evaporation rate to keep the film open and prevent skin drying.

FORMULATION

Compromise enters into the selection of primary film formers.

The Concept of Compromise

With the wide choice of primary and secondary film formers, plasticizers, and solvents, the formulator is presented with the problem of selecting the proper combinations for specific applications to paper coatings. To make this selection he must be guided by the following considerations:

    Use of the coated paper
    Desired properties
    Method of application
    Drying limitations
    Economics

For a given coating problem, a clear-cut, unequivocal solution in the way of formulation does not always exist. Every coating formula is a compromise between conflicting properties. A formula is never perfect in every respect; at best, it is an optimum compromise. However, so many variables are involved in the choice of ingredients, in the mechanics of coating, and in the properties of coatings that it is improbable—merely from a consideration of the laws of chance—that the formulator will arrive at the optimum coating as to quality and economy. Certainly, he is unlikely to hit upon the optimum solution of the problem without lengthy and exhaustive experimentation. The optimum formula therefore becomes a goal that is unlikely to be achieved except by a combination of perseverance, good judgment, experience, and luck. The formulator must often be satisfied with a solution that is workable, but one that he is not able to demonstrate to be the best. Undoubtedly, many materials used for coatings have not been studied sufficiently to determine whether they are the best available of the particular applications.

A few actual examples from practice help make clear the idea of compromise. The problem of determining the proportion of plasticizer to use in a given formula for a paper coating is the best illustration. If too little plasticizer is used, the folding properties are poor and the coated paper has a tendency to curl. If too much plasticizer is used, the coating is tacky and has a tendency to block. Seldom is there a definite range of plasticizer proportions where all three of these properties are the best possible. A compromise in which some sacrifice is made in folding and curl to obtain high blocking resistance is usually necessary. Another example is the problem of making a pigmented high-gloss coating. If the pigment proportion is increased, the hiding properties of the coating improve, but the gloss becomes poorer. A compromise must be made, therefore, between gloss and hiding power. This same problem of compromise is found in every phase of formulation. The formulator, therefore, must clearly understand the advantages, disadvantages, limitations, and effects of each of the substances that go into a coating.

Selection of Primary Film Former

In the classification of nonvolatile coating materials, an arbitrary distinction has been made between primary and secondary film formers. As the primary film formers constitute the basis of most solvent coatings for paper, it is helpful to make an analysis of their characteristics and to indicate as clearly as possible the applications where each one can be used to the best advantage.

Comparison of Cellulose Derivatives

Of all the cellulose derivatives, nitrocellulose has the largest application and offers the widest choice of formulation. Its greatest disadvantage is its high inflammability. Attempts in formulation to lower the inflammability of nitrocellulose coatings usually result in poor films with limited applications. The great advantages of nitrocellulose are its ready solubility in a wide range of solvents, its compatibility with many plasticizers and resins, its quick solvent release in drying, and its comparatively low cost. Because it was the first cellulose derivative to be used in coatings, the technique of its use has been extensively studied and developed over a period of years. As a result, it is a relatively simple material to handle, formulate, dissolve, and apply. In comparison with the other cellulose derivatives, nitrocellulose is usually found to have better adhesion to paper surfaces. From the economic point of view, it is cheaper than cellulose acetate, ethyl cellulose, and the mixed esters, on a volume basis. If the problem resolves itself into using cellulose derivatives, nitrocellulose is usually the answer, unless there is a compelling and clear-cut reason for using one of the other derivatives. Of course, cellulose acetate, acetobutyrate, acetopropionate, and ethyl cellulose, each has its special advantages, which for certain applications may outweigh the disadvantages of greater cost, more restricted range of formulation, and generally poorer adhesion.

Some examples of actual applications will help to clarify the process of making a choice.

If the problem is that of making a waterproof decorative coating whose main purpose is pleasing appearance, without any strict requirements as to low inflammability, resistance to sunlight and heat, greaseproofness, or low-temperature flexibility, one would in all probability choose nitrocellulose. For colored high-gloss or patent leather, matte, metallic, embossed simulated-leather, or crystal finishes for box coverings or for the printing trade, the outstanding advantages of nitrocellulose of lower cost, better adhesion, and easy formulation and handling would work in its favor.

For bronze metallic coatings, nitrocellulose has the disadvantage that it has a tendency to cause gelling in the solution after a period of several hours. Even here, however, it is preferred to the other cellulose derivatives, as the gelling problem can be overcome either by making smaller batches that can be used up within the gelling period, or by the use of gelling inhibitors.

For simulated leather finishes on kraft paper or on paper saturated with latex, nitrocellulose has all the advantages just mentioned and the additional one that it can be formulated to be embossed, without sticking, on a flat-bed press. The other cellulose derivatives are difficult to emboss on this type of press without sticking to the plate.

To make a coating of low water-vapor transmission, nitrocellulose offers better possibilities than any of the other cellulose derivatives. The incorporation of 1 % or less of paraffin wax into a properly plasticized nitrocellulose coating gives a film of low vapor transmission.

However, in spite of its outstanding advantages, nitrocellulose is far from a perfect coating meterial, and falls down on several counts. As mentioned already, its greatest drawback is its high inflammability, both in the raw state and in film form. Where it is required that the coated paper is not more inflammable than the paper itself, the formulator must forget nitrocellulose and turn to one of the other cellulose derivatives if he has decided for one reason or another to use this class of materials. Furthermore, nitrocellulose has poorer heat and light stability than the other cellulose esters and ethyl cellulose. It is not as resistant to some greases, chemicals, and food products as some of the other cellulose esters. It is also somewhat more difficult to formulate heat-sealing coatings with the R.S. type of nitrocellulose than with cellulose acetate and ethyl cellulose. Coatings of low-temperature flexibility can be made much more easily with ethyl cellulose than with nitrocellulose.

The preceding statements give some indications where cellulose derivatives other than nitrocellulose would be used to the best advantage. The exact choice depends on the fine judgment of the formulator. If the requirements call for a greaseproof, heat-sealing paper, and the choice is confined to the cellulose group, good judgment would seem to point to a choice of cellulose acetate or the mixed esters, rather than to nitrocellulose or ethyl cellulose. The latter is easy to formulate for heat sealing, but is the most soluble of the cellulose derivatives, and has comparatively poor grease resistance. In deciding between cellulose acetate and the mixed esters, several considerations must be weighed. Cellulose acetate is cheaper than the mixed esters and has somewhat better mechanical properties and grease resistance. However, it has a much narrower range of compatibility with solvents, resins, and plasticizers than the mixed esters.

Cellulose acetate also has higher water absorption and poorer moisture-vapor resistance than the mixed esters. The choice, therefore, will be a compromise.

Ethyl cellulose would be a logical selection where low-temperature flexibility or high resistance to alkalis is required. Other properties, acting in its favor, are solubility in cheap solvents, low specific gravity, low inflammability, good heat resistance, and better light resistance than nitrocellulose.

In connection with cost comparisons of film formers, it should be pointed out that the comparison should be made on a volume basis rather than on a weight basis. The significant cost figure is the *total* material cost per unit volume of nonvolatile coating material deposited. This includes both the cost of the nonvolatile materials and the cost of the required solvents.

Comparison of Vinyl Resins

Just as there is a wide choice for formulation in the cellulose-derivative group, there is also an equally wide choice in the vinyl-polymer group. Similarly, a clear cut selection is not always obvious and one must usually compromise. Representative members of the vinyl group already have been described in the section on primary film formers, but at this point, they will be compared with one another from the viewpoint of formulation and application. The following vinyl polymers will be considered:

Polyvinyl Acetate
Polyvinyl Chloride
Vinyl Chloride–Vinyl Acetate Copolymer
Vinyl Chloride–Vinylidene Chloride Copolymer
Vinylidene Chloride–Acrylonitrile Copolymer
Polyvinyl Butyral
Polyethylene

Each member of this group has its characteristic advantages, disadvantages, and limitations.

Polyvinyl Acetate is the most soluble member of this group, but even the polymers of higher molecular weight are thermoplastic, and blocking is always a serious consideration when this one is used. The block point can be raised by the addition of wax or nitrocellulose, but the film is never as good, from this viewpoint, as that of the higher molecular weight vinyl chloride–vinyl acetate copolymers. However, its thermoplasticity gives it good heat-sealing properties. Its greatest application is in the adhesive field and in pressure-sensitive coatings. Its freedom from odor, taste, and toxicity makes it useful for clear coatings in the food-packaging field, where it can be formulated with wax to reduce the blocking tendency. Pigmented decorative coatings, where the least blocking would spoil the appearance, and coatings to be embossed are definitely not in its field of application. Polyvinyl acetate is higher in cost on a volume basis than the grade of vinyl chloride-vinyl acetate copolymer commonly used in paper coatings.

Polyvinyl Chloride is restricted in solubility. Solutions of workable viscosities are almost always thixotropic and low in solids. However, films of vinyl chloride are much tougher and mechanically stronger than the films of high molecular weight vinyl acetate and are also much less thermoplastic. Its price on a volume basis is about the same as that of polyvinyl acetate, but the total cost per unit volume deposited, including solvent, is considerably higher because much lower solid concentrations must be used. Because of its limited solubility and its thixotropic behavior, the use of polyvinyl chloride is limited to applications where films of the highest mechanical strength and maximum resistance to blocking are required.

The Copolymers of Vinyl Chloride and Vinyl Acetate are a compromise between the wide solubility but high thermoplasticity of polyvinyl acetate and the restricted solubility but high mechanical strength and good blocking resistance of polyvinyl chloride. For most paper coatings, where vinyls are used, these copolymers are probably the most versatile choice. In common with polyvinyl chloride, however, they exhibit marked thixotropic and gel behavior—an important disadvantage in coating applications. Like all other vinyl resins, they also suffer from the disadvantage of high solvent retention.

The Vinyl Chloride–Vinylidene Chloride Copolymer, which is represented by the Geon 200 series, has similar properties to the polyvinyl chloride–vinyl acetate copolymer.[10] It is more soluble than polyvinyl chloride, but not as soluble as polyvinyl acetate. In comparison with equivalent molecular weight vinyl chloride–vinyl acetate copolymers, it is probably somewhat less soluble and has slightly greater tendency to gel in solutions of the same concentration. It is slightly higher in specific gravity than vinyl chloride–vinyl acetate copolymer.

The Vinylidene Chloride–Acrylonitrile Copolymer, as represented by Saran F-120, is soluble in a restricted range of solvents.[11] The only readily available true solvents for it are methyl ethyl ketone, cyclohexanone, isophorone, and mesityl oxide. Because methyl ethyl ketone is normally a readily available solvent, and because large proportions of it can be replaced with other ketones, esters, alcohols, and hydrocarbons, this restriction is not serious. With respect to solubility, it has the distinct advantage over the two copolymers already mentioned in that it does not have any tendency to gel in solutions of even high concentration. Therefore, it is much easier to handle in the coating operation. Another advantage of vinylidene chloride–acrylonitrile over the other two copolymers is that it has much lower moisture-vapor transmission. It appears to have the lowest moisture-vapor transmission properties in the vinyl group, with the possible exception of polyethylene. However, it has several marked disadvantages in comparison with the vinyl chloride–vinyl acetate and vinyl chloride–vinylidene chloride copolymers. It has greater solvent retention and generally requires more forced drying to eliminate this retained solvent. Furthermore, its specific gravity is about 18% higher than that of the vinyl chloride–vinyl acetate copolymer,

Polyvinyl Butyral can be dissolved in straight 95% ethyl alcohol, which is an advantage when the price of alcohol is low, but is no advantage today. In contrast with the vinyl chloride–vinyl acetate and vinyl chloride–vinylidene chloride copolymers, it exhibits no thixotropic or gel behavior and is consequently much easier to dissolve and to handle in solution. Another distinct advantage over the copolymers is that it contains no chlorine that can be split off as hydrochloric acid and deteriorate the film. It can be plasticized with large proportions of castor oil for some applications, which is also an advantage, since castor oil is cheaper than most chemical plasticizers. It has greater water absorption and higher moisture-vapor transmission and probably has more tendency to block in the uncured state than the copolymers. It has the lowest specific gravity of any of the vinyl resins except polyethylene, but that advantage is more than balanced by its much higher cost. Compared with the vinyl chloride–vinyl acetate copolymer, its cost on a volume basis is more than three times as high. It probably has slightly higher solvent retention than the copolymer.

Polyvinyl butyral can be formulated with certain thermosetting resins so that it can be baked at elevated temperatures to form a cured or "vulcanized" film, that is insoluble and nonblocking. Although this type of formulation has a great many applications for fabrics, its use for paper is limited because of the baking cycle required.

Polyethylene offers many advantages over other vinyl resins. It has a lower specific gravity than any other member of either the vinyl or the cellulose group. Films of polyethylene have a greater temperature range of flexibility than those of the other vinyl derivatives. It is unaffected by any organic solvent at room temperature. It requires no plasticizer, and therefore its films are free from migration troubles. It has less solvent retention than the other vinyl compounds and probably as low moisture-vapor transmission as the vinylidene chloride–acrylonitrile copolymer. Up to this point, this material seems almost too good to be true. It is not surprising, therefore, that it has several distinct drawbacks. Its chief disadvantage is that it is insoluble in any organic solvent at room temperature and it must be dissolved and used in hot solutions. It is more inflammable than the vinyl chloride–vinyl acetate copolymer. It is also difficult to formulate for high-gloss coatings. Another disadvantage is that it cannot be heat-sealed with electronic equipment. It can be used to the best advantage for highly resistant, highly flexible, low moisture-transmission coatings, where the coater is equipped to handle hot solutions and is prepared to cope with the additional hazards that their use entails.

Vinyl Resins vs. Cellulose Derivatives

A great deal of discussion in the coating industry today concerns the subject of vinyl vs. cellulose derivatives. The particular cellulose derivative that is usually considered in this comparison is nitrocellulose. Much confusion of thought occurs and opinions are often conflicting.

In a survey in which users of both vinyls and nitrocellulose were interviewed, the results show more than a lack of unanimity of opinion with regard to preference for particular applications. This problem of evaluation and comparison is approached here with a knowledge that the process of comparison is still going on, and that the last word on either side has not yet been said.

Certain general statements can be made. First, much tougher and more flexible coatings can be obtained with some of the vinyl resins than with any cellulose derivative. This is an advantage in favor of the vinyls. However, the vinyl resins are considerably slower drying than the cellulose derivatives, because of their much greater solvent retention. Furthermore, they must be dried at comparatively high temperatures to remove this retained solvent and to attain maximum adhesion. This is a marked disadvantage of vinyl resins. Some coating manufacturers have drying equipment that is adequate for cellulose derivatives, but not for vinyls.

These two bases of comparison–mechanical properties and speed of drying—apply as a *general* comparison between the cellulose derivatives and the vinyl resins. In comparing mechanical properties, the equivalent viscosity types of each group should be considered. In other words, it is unfair to compare low-viscosity vinyl acetate with high-viscosity nitrocellulose.

Because the copolymers of vinyl chloride and vinyl acetate are the most generally used types of the vinyl group and nitrocellulose is the most generally used derivative of the cellulose group, it will simplify the discussion to use these two film formers as the basis for further comparison. The other members in each group have been compared with one another, and their positions in this discussion can be interpolated from the relationships of their properties with these two representative plastics.

For applications where maximum toughness, flexibility, good folding properties, and abrasion resistance are desired, the copolymers are superior to nitrocellulose. For example, for coating catalog covers, which are subjected to considerable handling, the copolymers would be the logical choice, bearing in mind, of course, their disadvantage of slow drying. The other advantages are better resistance to strong acids and alkalis, alcohol, greases, and oils, noninflammability, and easier heat-sealing properties. These properties make them more suitable for most food and drug packaging. Where maximum toughness and noninflammability are not needed, as in most decorative papers, easier formulation, easier handling, and quicker drying make nitrocellulose more desirable.

General statements are dangerous to make, but very broadly speaking, nitrocellulose is probably more applicable in the decorative paper field, whereas the vinyl copolymers have greater advantages for functional papers.

Chlorinated Rubber and Cyclized Rubber Compared with Other Film Formers

By comparison with nitrocellulose, chlorinated rubber is noninflammable; it is soluble in much cheaper solvent combinations; and it has better resistance to strong acids and alkalis. However, it has poorer tensile strength and flexibility; it is more thermoplastic and has more tendency to block; it has poorer resistance to oils and greases; and it has more solvent retention, with resulting slower drying than nitrocellulose.

Cyclized rubber is soluble in cheap hydrocarbon solvents, both aliphatic and aromatic. Compared with nitrocellulose, it has much better heat-sealing properties, better moisture-vapor resistance, and better acid, alkali, and alcohol resistance. Its films can be made noninflammable. Disadvantages with respect to nitrocellulose are slower drying, greater tendency to block, tendency to oxidize, poorer mechanical strength, and much less resistance to oils.

Compared with the vinyl chloride–vinyl acetate copolymer, cyclized rubber is soluble in much cheaper solvent mixtures; it is free from thixotropic or gel behavior; it has better moisture vapor resistance; it contains no chlorine to split off as HCl. Cyclized rubber has much poorer mechanical properties and much less resistance to oils than the copolymer.

Compared with chlorinated rubber, cyclized rubber is soluble in cheaper solvents. It has somewhat better moisture-vapor resistance and heat-sealing properties and does not contain chlorine. It has much poorer general resistance to oils and greases.

The outstanding property of cyclized rubber is its high moisture-vapor resistance. Available data indicate that per volume of film deposited it has the lowest transmission rate among the well-known film formers, and that for a given vapor transmission rate thinner films can be used. Therefore, it is most suitable for food and drug packaging, where this property is very important.

Choice of Viscosity

When the formulator, according to his best judgment, has selected the primary film former for a particular application, he is often confronted with the further problem of selecting the proper viscosity or molecular weight. As in the selection of the primary film former, the choice of the viscosity will also be a compromise. Low-viscosity polymers have the advantage of giving higher solid concentrations at the working viscosity and depositing thicker films, with less solvent consumption per unit volume of dry film. Low-viscosity types are therefore more economical to use. Higher-viscosity types of any polymer series have better tensile strength, flexibility, folding, and abrasion resistance. In the case of highly thermoplastic compounds, such as the vinyl resins, the lower molecular weight members are much more thermoplastic and have lower softening points, i.e., more tendency to block. The lower-viscosity vinyls are easier to dissolve and are not as subject to thixotropic and gel behavior as the higher-viscosity members of the same series. The optimum compromise for paper coatings in the vinyl chloride–vinyl acetate copolymer series is probably the VYHH-1 viscosity grade.

A practical example helps to make this descussion clear. Assume that nitrocellulose is to be used as a primary film former in a pigmented coating for a 0.020-in. paper stock. The principal requirements are high gloss, good hiding, and good folding properties. Assume that ½-sec nitrocellulose, properly pigmented and plasticized, gives a coating of excellent gloss, but that the folding properties are not as good as desired. In addition 5—6-sec nitrocellulose, properly pigmented and plasticized, has satisfsctory folding properties, but the gloss and hiding power are inadequate because of the thinner film deposited from the lower solids concentration.

If this coating is pigmented to a greater extent to improve its hiding, its degree of gloss is further reduced and its folding becomes poorer. Of course, a thicker coating can be applied, but assume for simplicity that the coating machine is applying the maximum thickness of wet coating in each case. Even if a thicker wet film can be applied, the higher-viscosity nitrocellulose coating would be more expensive per unit of dry volume deposited, for a lower nonvolatile concentration must be used. The best solution to this problem is the use of a mixture of $\frac{1}{2}$-sec nitrocellulose and 5—6-sec nitrocellulose, which represents a compromise between gloss and hiding on the one hand and good folding on the other.

In general, for applications where good mechanical properties are desired, the higher-viscosity polymers are used. In applications where decorative properties are the chief aim, lower-viscosity types usually serve to the best advantage.

Plasticizers and Other Agents Used in Formulation

The use of resins and plasticizers has been discussed to some extent in the previous sections. In selecting plasticizers, the chief consideration is to obtain a coating of good elasticity, folding, and flatness without sacrificing the inherent good properties of the primary film former, such as grease resistance, water resistance, moisture-vapor resistance, tensile strength, elasticity, and abrasion resistance. This must be accomplished by a proportion of plasticizer short of the point where the film is tacky. The selection of plasticizer and the amount used will be a compromise.

The formulator should be cautious with the addition of resins. This statement of course does not apply to the soft resins, which are very useful as plasticizers, nor to the vinyls and other resins which have been classified here as primary film formers. Resins have much more applicability for rigid surfaces such as wood and metal than for flexible surfaces such as paper. The reasons for the use of resins, as well as their disadvantages, have already been discussed. In general, the softer types are more useful than the hard brittle resins.

Other substances that enter into coating formulas and which have not been discussed are waxes and stabilizers. Waxes are used to improve blocking resistance and, sometimes, moisture resistance. Stabilizers are often used to improve heat and light stability. They are generally incorporated with vinyl resins and, sometimes, with nitrocellulose and ethyl cellulose.

BASE PAPER

The type of base paper used for coating depends on the end use of coated paper. Many types of papers are coated; they vary in fiber composition, thickness, sizing, mechanical strength, and surface finish. They include many types of kraft, sulfite, and soda stocks; glassine;

board and cover stocks; and special saturated papers. The characteristics of the base paper have an important bearing on the formulation of the coating, the type of coating equipment used, and the final results obtained.

The mechanical characteristics of the paper, such as tensile strength, bursting strength, tear resistance, and folding properties, should be adequate both for its end use and for its ability to pass through the coating operation satisfactorily.

The finish of the paper is important, because it determines to a great extent the appearance of the finished coating and its functional characteristics. A surface that is highly absorptive requires more coating than a relatively nonabsorptive surface. However, the adhesion of the coating to the absorptive surface is likely to be better.

For applications where smoothness and continuity of the coating are important, the base paper is often supercalendered or precoated with an aqueous size, such as casein. This provides a good surface for the solvent coating. It is cheaper to precoat a rough paper with an aqueous sizing coat than to use extra coats of lacquer to obtain the desired results. The precoating should adhere well to the base paper and should not dust; it also should be flexible enough to withstand a sharp fold. It is sometimes more difficult to obtain good adhesion of the lacquer to a precoated sheet and this possibility should always be considered.

Mechanical strength must often be sacrificed to obtain a paper with a smooth surface for coating, particularly with certain types of papers that are to be solvent coated for decorative effects.

Heavier papers, such as cover stocks, should have satisfactory folding properties, so that when the coated stock is folded the paper will not crack the coating.

Large quantities of paper are impregnated before coating. Absorbent paper may be impregnated with natural rubber latex, synthetic rubber latex, resins, or glue and glycerin to produce tough papers that are flexible and that have high tear resistance. These impregnated papers, when coated, are often used as artificial leather. Impregnated stocks present special problems in adhesion and in the possibility of reaction with the solvent coating. Rubber-impregnated bases may require special bonding coats to secure proper adhesion. Papers impregnated with plasticized resins present the possibility of plasticizer migration from the paper to the solvent coating with resulting tackiness. Glue-glycerin impregnations sometimes "sweat" through the solvent coating in humid weather, because of high moisture pick up.

Kraft paper is sometimes pretreated or plasticized with a glycerin solution before coating to improve its flexiblity and folding properties.

Difficulties are often encountered in the coating operation as a result of poor mechanical condition of the base stock, because of improper processing, winding, and calendering in the paper mill. Among the the possible defects are untrimmed and rough edges, poor splices, tears, holes, wrinkles, and hills and valleys. Wrinkles and hills and valleys often cause serious difficulties in obtaining uniform coverage, particularly when colored coatings are used.

COMMON COATING TROUBLES

Streaks in a colored coating can be traced to insufficient mixing or to selective settling after mixing. Excessive viscosity, which results in uneven flow on the coating machine, is also a possible source of this defect.

Light Spots may occur in coatings made on supported knife-type coating machines, by lacquer or foreign matter becoming deposited on the supporting roll, plate, or belt and causing excessive pressure of the paper against the knife. They may also occur on coatings made on roll coaters and print coaters when foreign matter collects on the rolls.

Specks in the coating are due to improper filtration of the coating solution or to seeding of pigmented lacquers after filtration, especially if the lacquer is stored for some time before use. When specks are found in the coating, the lacquer should always be refiltered.

Scratches are based by particles collecting under the knife or on the rolls of coaters. They are more likely to occur on knife-type coaters, because of the small clearances involved. These particles may come from the lacquer, the paper, or the air.

Bubbles may be caused by air entrapped in the solution, by too-rapid drying with excessive temperatures, and by pinholes in the paper that release air when the coating solution is applied.

Orange Peel is the hill-and-valley effect resulting from poor solvent balance or from too-rapid drying. It may be corrected by reducing the speed of drying with slower evaporating solvents, or by increasing the proportion of true solvent in the solution.

Blushing has been discussed under "Solvents." It may be caused by condensation of moisture from the air, or by precipitation of some part of the nonvolatile by an upset of the solvent balance at some point in the drying process, which creates an excessive proportion of diluent.

Water blush can be corrected by reducing the water-miscible solvent content of the solution, increasing the proportion of slowly-evaporating solvents, increasing the temperature, or dehuminifying the air. A diluent blush may be corrected by increasing the proportion of true solvent in the formula.

Curling may be caused by low moisture content in the paper. Another cause is insufficient plasticizer in the formula, which results in excessive shrinkage on drying.

Blocking is the term applied to the sticking of the coated side to the adjacent uncoated side in a roll of paper or in a pile of sheets. It is caused by tackiness resulting either from solvent retention or from an overplasticized film. If caused by solvent retention, it is often difficult to correct, and modification of the formula or of the drying method may be necessary.

A coating that is apparently block resistant under ordinary conditions is sometimes found to block in certain portions of a tightly wound roll. This may happen when a roll is wound unevenly and excessive pressure develops in a ridge. Little information is available on pressures in unevenly wound rolls, and this should be an interesting subject of investigation.

GENERAL ECONOMICS OF SOLVENT COATINGS

The cost of a solvent coating is made up of the cost of the coating solution and the cost of its application.

Cost of the Coating Solution

The cost of the solution per unit area of paper depends on the amount of coating necessary to do the job satisfactorily. This figure is not always easy to determine, and involves a study of the relationship between the thickness of the deposited film and the properties of the coated paper. Properties like gloss, hiding, folding, moisture-vapor resistance, heat sealing, and grease resistance improve as the thickness of the film is increased. Speed of drying, blocking resistance, and cost are unfavorably affected by increased film thickness. Determination of the optimum thickness becomes a matter of good judgment.

After determining the required thickness of the coating, the aim should be to use a coating solution that will have the lowest cost per unit volume of dry coating deposited, consistent with the desired quality of the coating and with the mechanics of application. The cost depends on the materials used and on the solids concentration of the solution. Both factors can be controlled by the formulator within cer-

tain limits. The cost of the coating can be lowered by loading the film
with cheap resins, plasticizers, and fillers. The limitations of this procedure with respect to quality are obvious. Higher solids concentration for the same viscosity can be obtained by using lower molecular
weight grades of primary film formers, but this value is limited by considerations of the mechanical strength of the film and in some cases by
the increased possibility of blocking with lower molecular weight and
lower softening point polymers. Higher solids concentrations can
usually be obtained by using fast-evaporating solvents, which produce
lower viscosities. This is limited, however, by considerations of good
flow and by the necessity for depositing films free from the defects of
too-rapid drying. Finally, solids concentrations can be increased at
the expense of viscosity. These values are limited by the optimum
workable viscosity required by the method of application.

The cost of the solvent mixture depends on the types of solvents
used and on the quantity of cheap diluents that can be blended into
the formula. Currently, the costs of ester solvents and ethyl alcohol
are high, while the costs of the lower ketones, isopropyl alcohol, and
methanol are condiderably lower. The costs of aromatic and petroleum
hydrocarbons are usually much lower than those of the other commonly
used solvents. Good economics require one to use as much cheap
hydrocarbon as possible in the solvent mixture. For solutions of primary film formers, where the hydrocarbons are diluents, the proportion to be used is limited by considerations of tolerance and viscosity.
As the proportion of diluent is increased, cost of the solvent mixture is
lowered, but a point is reached where lower solids concentration must
be used to keep within the working viscosity. The proportion of diluent, therefore, should be an optimum compromise between solvent cost
and solids concentration.

Cost of Application

Equally important in the cost of coating is the direct cost of application of the coating solution to the paper as determined by the speed
at which the paper is coated, the number of coats applied, and the
width of the paper.

The speed of coating depends on the coating machine, the type of
paper, and the limitations of drying. The drying speed depends on
the amount of coating applied, the evaporation rate of the solvent, the
solvent retention of the coating, and the drying equipment. The number of coats depends on the required thickness of the coating, the con-

centration of solids in the solution, and the amount of coating the coating machine can apply in one operation. The speed of drying and the amount of solution applied per coat are often conflicting. In multiple-coat operations, if the amount of solution applied per coat can be increased without decreasing the speed of drying proportionately, it will be economical to do so, provided that one coat is saved in this way. The saving of one coat does not always result, as the following illustration shows.

Let us assume that a coating with a dry thickness of 0.0020 in. is desired and that the coating machine ordinarily deposits a film of 0.0005 in. thickness per coat. Under ordinary conditions, four coats would be required to attain this thickness. Unless the thickness per coat can be increased to at least 0.0067 in., four coats will still be required. If the coating machine can apply a maximum thickness of only 0.0006 in., no reduction in the number of coats results even though 20% more coating is applied in the first three coats. Table 5-60 shows what happens when the speed must be reduced proportionately because of drying limitations.

TABLE 5-60

**Effect of Coating Thickness on the Required Drying Time
of Solvent Coating**

Basis=1,000 yards of paper per roll; the speed is inversely proportional to the film thickness deposited, and equals 50 yards per minute when the film thickness is 0.0005 in.

	Case 1		Case 2		Case 3	
Coats	Thickness per coat, in.	Coating time per coat, min.	Thickness per coat, in.	Coating time per coat, min.	Thickness per coat, in.	Coating time per coat, min.
1	0.0005	20	0.0006	24	0.00067	26.7
2	0.0005	20	0.0006	24	0.00067	26.7
3	0.0005	20	0.0006	24	0.00067	26.7
4	0.0005	20	0.0002	8		
	0.0020	80	0.0020	80	0.0020	80.1

In multiple-coat operations, if the speed is inversely proportional to the amount of coating deposited, no saving in actual coating time is effected even though one coat is eliminated. However, the elimination of one coat saves the time consumed in changing rolls from one coat to

the next and it also reduces waste of paper, which is likely to be encountered in multiple-coat operations. One must keep in mind also the limit
in film thickness that can be deposited in one coat. This value may be
short of the limits imposed by the coating and drying equipment. It
is the maximum thickness that can be deposited without obtaining
bubbles, skin drying, streaks, and excessive retained solvent.

The preceding example is an extreme case used merely to bring out
a point. Most coating operations involve only one or two coats. One
exception is coated rubberized paper used for artificial leather, which
is often given three to five coats, depending on the results desired.

Fig. 5-1. Manually operated two-adsorber recovery plant (*Courtesy of
Pyrotex Leather Co.*).

SOLVENT RECOVERY

If the amount of solvent used is sufficiently large, the recovery of the
solvent vapors from the exhaust air of the coating machines may result
in important savings and may be a decisive factor in determining
whether the coating plant operates at a profit or at a loss. Solvent

recovery also reduces fire and toxicity hazards by eliminating the discharge of solvent vapors into the atmosphere. Almost any organic solvent or combination of solvents, including alcohols, esters, ketones, hydrocarbons, and chlorinated compounds, can be recovered in a properly designed recovery plant.

Two general methods of solvent recovery are in use: adsorption and condensation. For the solvents generally used in coatings, adsorption by activated carbon has proved to be the more effective method, and nearly all installations in this field are activated-carbon systems. The following discussion is confined to this method.

A recovery system consists of the following basic units of equipment: a fan system for exhausting the vapor-laden air from the driers through ducts leading to the recovery plant; cooling coils for lowering the temperature of the hot exhaust air before it is passed through the carbon; two or more adsorbers containing the activated carbon; a condenser for the solvent vapor and the steam used to distill the vapor from the carbon; an automatic decanter for separating the solvent and water layers, or settling tanks, where these layers separate by gravity; a distillation system consisting of a still pot, a fractionating column, and a condenser for removing the water and for separating the solvents; process and storage tanks; pumps for transporting process solvents from the storage tanks to the still and for returning recovered solvents to the mixing room; control, indicating, recording, and safety instruments.

A recovery plant can be designed for either manual or automatic operation. Automatic is warranted for large installations, but the initial investment is much higher than for a manually operated plant.

A two-adsorber, manually operated recovery plant (Figure 5-1) works as follows:

The vapor-laden air is exhausted from the driers by the fan system and is passed through a replaceable filter to remove lint and other suspended matter. It then goes through a cooling coil unit, which lowers its temperature for more efficient adsorption. The fan may be placed between the cooler and the adsorbers and the air blown through the carbon under positive pressure; or it may be placed after the adsorbers and the air may be drawn through under negative pressure. The second method has the advantage that any leak in the system is air leakage inward, rather than vapor-laden air outward. The first method has the advantage that the fan always handles cool air rather than hot air.

From the cooling coils, the vapor-laden air goes into adsorbers and passes through the carbon beds, which are supported on grates. When one adsorber becomes saturated, it is shut off, and all the air is passed through the other adsorber. The point of saturation is determined from a knowledge of the rate of solvent evaporation on the coating machines and of the adsorption capacity of the carbon; or it is determined by a continuous check of the air from the adsorbers, with a sensitive vapor detector, for the first signs of the presence of solvent vapor. Steam is then admitted to the closed adsorber and the adsorbed solvent is distilled from the carbon at atmospheric or elevated pressure, depending on the nature of the solvents used. The steaming operation requires 15—60 minutes, depending on the types of solvents used and on the characteristics of the adsorber.

When almost all the solvent has been distilled off from the adsorber, as determined by checking the specific gravity of the condensate, the steam is turned off and the air inlet and outlet of the adsorber are opened. When the second adsorber is saturated, the cycle is repeated.

This adsorption operation may be modified by including a period after the steaming operation in which the carbon is partially cooled and dried. In this procedure, the vapor-laden air is passed through the first adsorber and is then discharged into the second by an additional blower, where it dries and cools the carbon that was steamed previously. As soon as the exhaust air from the first adsorber contains an appreciable amount of solvent, the air flow is switched to the second adsorber and the first adsorber is shut off and steamed. The cycle is then repeated.

The distilled solvent and water vapor from the adsorbers are passed through a condenser, and the condensed liquid is allowed to fall into an automatic decanter, which separates the layers; or it is pumped into setting tanks for the same purpose. In an automatic decanter, the separated layers usually drop by gravity to process storage tanks.

The top layer contains the water-immiscible solvents, such as hydrocarbons, most of the ester solvents, a small portion of the alcohols and soluble ketones, and a small amount of water. The bottom layer consists of water and most of the water-soluble solvents, like alcohols and acetone, and part of the ester solvents. These two layers are pumped separately into the still pot and distilled to remove the water and to separate the solvents by fractionation, if necessary. The top layer usually is given a simple distillation in which a fraction containing all the water is distilled over. This initial fraction is run back into the decanter, while the water-free portion remaining in the still is distilled into a finished solvent tank. The bottom layer is distilled through a fractionating column to remove the large amount of water it contains and the distillate sent to either the recovered-solvent tank or to a processing tank for further treatment. This may consist of distillation over calcium chloride to remove any excess water.

The dry solvents may be caught in a common finished solvent tank or they may be fractionated to separate the constituent solvents. The former method is much simpler and it is suitable where the same mixture can be reused or by blending with fresh solvents. However, where radically different solvent formulas are employed for different types of coatings, and the mixture cannot be used in its entirety some separation must be made. This process is often difficult, particularly if the mixtures are azeotropic.

DRYING

As stated early in this chapter, the drying rate of commonly used solvent coatings is considerably higher than that of aqueous coatings. This high drying rate presents special problems that must be taken into consideration. With respect to external conditions, the rate of drying is a function of the difference between the temperature of the surrounding air and that of the coating and also of the air velocity past the paper.

An upper limit to both these factors exists. Excessive temperature differences and air velocities may cause bubbles, skin drying, and orange peel. On the other hand, low air temperatures sometimes promote blushing, especially in high humidity. The proper drying conditions depend on the nature of the solvent mixture and must be determined by experiment.

Solvent coatings should always be dried in a closed chamber. The required length of this chamber depends on several factors: the speed, the thickness of the coating, the nature of the solvents, the temperature inside the drier, and the air movement. In some driers, the paper goes to the end and comes back one or more times to take maximum advantage of the drying space. In general, however, if sufficient length is available, it is more satisfactory to make only one pass through the drier. This procedure decreases the possibility of breaks, tears, and curling, and facilitates inspection and control. In practice, drier lengths vary from 20 to 200 ft.

Heat to the drier may be supplied by several methods: hot air, steam coils, or infrared lamps. Hot air is generally the most satisfactory method. Steam coils alone, without controlled air movement, often work in many cases, but they are much less effective than a well designed hot-air system. Infrared lamps are advantageous when high temperatures, which cannot be obtained with air heated by reasonable steam pressure, are desired. However, they have not been available in explosionproof construction, and are therefore not considered completely safe where high solvent concentrations are likely to be encountered.

In a hot-air system, common practice is to heat the air by passing it through fin-type steam heated coils from which it is sent through a duct system to the drying chamber. The hot air must enter the drier in such a way as to obtain the best possible distribution, to avoid localized, overheated areas. A very effective practice is the use of a flat duct inside the drier, extending over the entire width of the chamber with perforations on the bottom side through which the air is directed toward the coated surface. The bottom of the duct may be placed at 6—12 in. from the surface of the paper. Usually a portion of this air is recirculated and the rest exhausted by a separate fan. This duct should start some distance from the front of the drying chamber, because the coating in the initial stages is wet and blistering and skin drying may result from air impinging on the wet film at high velocity. Good practice requires comparatively low temperatures at the beginning, with a gradually-rising temperature gradient. A gradual temperature gradient is very difficult to obtain with a single-fan and hot-air system, and it is often advisable to have two or more independent fans and hot-air supplies, each with separate controls.

Sufficient air must be exhausted from the drier by an independent fan to keep the solvent concentration well below the lower explosive limit. The exhaust air should be taken from more than one point to make sure that no localized pockets of high solvent concentrations exist.

One of the most important problems, in the drying of solvent coatings, is solvent retention. Many types of coatings, particularly those made with vinyl resins, have a tendency to retain solvent. Cellulose derivative coatings with low resin content have this tendency to a much lower degree. The retained solvent is sometimes very difficult to remove and may result in the "blocking" of the coating when the paper is wound. Experience has shown than gradual drying is more effective than very rapid drying, in the initial stages, in overcoming this trouble. High-temperature forced drying is often necessary to remove this solvent. Sometimes, however, high-temperature forced drying, for a short period of time, seems to aggravate the trouble. Festooning for 30 minutes to 1 hour after the paper emerges from the drier, at fairly low temperature with good air movement, is often effective. Each formulation presents its own particular drying problem, and the drying method for eliminating the retained solvent should be studied in each case.

MIXING, GRINDING, AND FILTRATION

Mixing, grinding, and filtration are unit operations that play important parts in preparing solvent-coating solutions.

Mixing

In making solvent-coating solutions, three distinct dispersion problems arise: (1) dissolving the soluble materials; (2) dispersing dry colors and fillers by grinding; and (3) incorporating the ground materials into the solution by agitation

Many types of mixing equipment are suitable for dissolving and mixing: propeller; turbine; paddle; gate; screw; pony; ribbon; rotating drum; kneader; and special types, such as the Abbé Lenart and Cowles mixers.

The type of mixer selected depends on the size of the batches, the type of solution, the viscosity, and the time limitation for making the batches. Basically, mixing is a simple operation that, reduced to its crudest form, can be done with a paddle and a bucket. To perform the operation efficiently, however, careful consideration must be given to the factors just mentioned in selecting the type of mixer to be used.

For solutions and mixtures up to about 4,000 cP viscosity, the types most commonly used are the propeller, turbine, paddle, and gate. The last three are also suitable for higher viscosities, as are the pony, Abbe Lenart, and Cowles mixers. The kneader mixer is used for very viscous solutions and thick pastes. Propeller mixers usually operate at 400—1725 rpm. Turbine mixers run, as a rule, at peripheral speeds of 600—1,800 ft/min, depending on the diameter of the turbine and on the height of the solution to be mixed. Paddle and gate mixers are run ordinarily at speeds of 30—100 rpm.

Tanks for dissolving and mixing large batches, which must be cleaned only rarely, should be fitted with permanently fixed, tight covers to reduce the loss of volatile solvents to a minimum during the mixing operation. The cover should be equipped with an adequate manhole. Better practice is to use an independent, direct, completely enclosed gear-reduced drive for each tank rather than to drive a series of tanks by belts from a single motor.

Where two floors are available, one should set the tank on the lower floor in such a way that the top comes up through the next floor. The solid components can then be easily dropped into the tank and the solvent can be pumped, or drained from drums, into the tank opening. The tank opening should be small enough, or so constructed, that it is impossible for anyone to fall into it accidentally.

Solvents in large quantities can be handled conveniently by pumping from storage tanks through carefully calibrated meters directly into the mixing tank. Good practice requires individual pumps and meters for each solvent, to prevent accidental contamination.

In a coating plant where many small batches of different colors are used, there is always the problem of changing colors with a minimum

Fig. 5-2. Portable mixing unit, including lifter, truck, container, and agitator (*Courtesy of Pyrotex Leather Co.*).

of cleaning and washing of mixing tanks. To solve this problem, one should make the clear base solutions in large mixing tanks (Figure 5-2) pumping them, as needed, to individual small mixing containers of 50—100 gal capacity, where pigment pastes, fillers, and other ingredients can be added. Enough small mixing units should be available that shades of similar color can be mixed in the same container. One very convenient method of handling these small batches is to use portable mixing containers of 50—100 gal capacity, with portable agitators mounted on four-wheeled trucks. These units, including truck, mixing container, and agitator, can be wheeled to a platform scale for loading with clear solution, pigments and other materials. The units can be designed so that they can be wheeled to the coating machine and raised by a hydraulically or mechanically-operated lifter above the coating machine. The solution can then be run by gravity directly to the coating head. In this way, the solution can be continuously agitated (to prevent settling,) while it is being used.

If possible, in changing from one color to another, one should start with a batch of a light shade and use succeedingly darker shades for the following batches in the same mixing container. When the darkest shade is reached, some form of cleaning operation is necessary before the container can be used for a light shade again.

Grinding

When pigments and fillers are used, they always must be wet with plasticizers or the clear coating solution and ground to obtain fine particle size. Several types of grinding equipment can be used. Among these are pebble and ball mills, roller mills, and stone mills.

When pigments are to be ground in a solvent solution, pebble or ball mills are advantageous, because the grinding operation is done in a closed system and the solvent loss is at a minimum. Pebble and ball mills, however, require a long grinding time—20 hours or more—and usually produce grinds that give lower gloss than those obtained by other types of grinders.

Stone mills produce grinds of high gloss, but they have low grinding rates and require frequent dressing of the stones for best results. They are not ideally suited for grinds in solvent solutions, because of the high evaporation loss.

Roller mills are very suitable for grinding in nonvolatile plasticizer pastes. They have good production capacity and produce grinds of good gloss and smoothness. The fineness of grind depends on the closeness of the setting, the number of passes, and the type of pigment. On a three-roll mill, several passes are usually necessary. Grinds made on a five-roll mill require fewer passes than those made on a three-roll mill.

Filtration

The solution should always be filtered before coating to remove large undispersed particles, iron rust from drums, and other foreign matter. Filtering clear solutions does not present a serious problem, and different filters can be used, such as plate-and frame-type filter presses, centrifugal clarifiers, metal-screen filters, or ordinary cloth bags. However, the filtration of colored solutions, where many batches of different colors are used, is more difficult, because the use of a permanent filter involves a difficult cleaning operation each time the color is changed. Removable cloth bags tied to the outlet of the mixing container are often satisfactory if the viscosity is not too high for rapid flow. High-viscosity mixtures can be pumped through a removable screen or cloth filter. However, this procedure always imposes the problem of cleaning the pump and pipe lines when color changes are made.

COATING METHODS AND MACHINES

An ideal coating machine or process for the application of solvent coatings to paper should have certain general characteristics, outlined as follows:

The ability to apply a film of lacquer of the desired thickness uniformly across the sheet, free from streaks, scratches, and other mechanical blemishes;

A quick and accurate method of adjustment, so that the amount of lacquer applied can be varied from a fraction of 1/1000 in. to 2/1000 in. or more of finished thickness, with a precision of at least 10%;

A minimum tendency to cause breaks in the web of paper as it passes through the narrowest constriction of the coating head;

The ability to be cleaned quickly and easily between colors, or after a break in the web;

The ability to coat any width up to the maximum with a minimum of manipulation and adjustment;

Low exposure of the bank or bath of lacquer to the open air, to keep loss of solvent by evaporation to a minimum;

The ability to apply coating solutions of a wide range of viscosities;

A means for delivering an uninterrupted supply of lacquer to the coating head;

Freedom from excessive vibration or gear chatter, which might have a tendency to produce vibration lines in the coating.

No single coating machine possesses all these characteristics and and every type used suffers from one deficiency or another.

The basic types of coating machine are few in number and simple in principle. The following types will be described briefly: knife coaters roll coaters, spray coaters, air-brush coaters, and print coaters.

The most widely used types for solvent coatings are knife coaters and roll coaters.

Knife Coaters

Several variations of the knife coater are commonly used. Each has its own advantages and disadvantages.

Suspended- or Floating-Knife Coater

This consists of a knife against the face of the paper and has no support directly underneath. The coating solution is fed in front of the knife and lies on the paper in a puddle or bank. Supporting rolls or bars are in front and in back of the knife over which the paper passes. The lacquer is spread by the knife and the excess drops over both edges of the paper, where it is caught in a receiver and returned to the main lacquer supply. The amount of coating deposited is determined by the tension of the paper, the pressure of the knife against the paper, the angle at which the knife is set, and the thickness and shape of the bottom face of the knife. The ends of the knife are supported in trunions fitted with micrometer screws for up or down adjustment. The angle of the knife is also adjustable. This type of knife coater was originally designed for coating cloth with thick coatings of high viscosities. In the paper field its widest application today is for coating heavy, saturated paper bases with high-viscosity solutions for making artificial leather. The knife coater is not suitable for applying smooth, uniform coatings to light-weight papers, cover stocks, or board stocks because of its relatively poor control. Its chief advantages lie in its simple construction, ease of cleaning, and its ability to coat without a margin on each side.

Coater with Knife Against Paper Roll

Partly to overcome the lack of positive control inherent in the floating knife, the arrangement is often modified by introducing a supporting roll, 8 in. or more in diameter, directly under the knife. The web of paper wraps around this roll, which is usually driven at the same surface speed as that of the paper. Because of the positive adjustable clearance between the face of the knife and the supporting roll, this type of knife coater lends itself to much closer control than the floating knife. The closeness of control depends on the accuracy with which the face of the knife and the surface of the roll have been machined; it depends also on the fineness and accuracy of the micrometer screws used to adjust the knife up or down;finally, it depends on the uniformity of the thickness of the paper, and its freedom from wrinkles, hills, and valleys.

The roll may be made from metal or from some type of rubber. Each material has its advantages and disadvantages. A metal roll can be machined more accurately and its accuracy maintained over a longer period of time than a for rubber roll. A rubber roll, however, has resiliency and it may be better adapted for paper that is likely to have poor uniformity of thickness. It is also better adapted to handle poorly made splices. On the whole, for smooth, uniform light-weight papers, a metal roll is preferable. A steel roll is the best for most purposes. Its chief disadvantages are its susceptibility to corrosion and the possibility of its sparking when struck with a ferrous object. Stainless steel overcomes the corrosion tendency, but it does not do away with the sparking hazard. Bronze and aluminum will not corrode or spark, but they are much more easily chipped, dented, and scratched.

The coating solution flows onto the paper, in front of the coating knife, from a reservoir or drum. Adjustable guides are used to confine the lacquer on the paper and to prevent its running to the supporting roll. In some designs these guides take the form of a completely enclosed box that acts as a confining reservoir for the lacquer. This box has a slot at the bottom, which is usually adjustable, to obtain the the required width of coating. In a recent modification, the box is replaced by a tube, which is permanently attached to the knife. The lacquer is introduced into an inlet to the tube and flows out through an adjustable slot at the bottom.

The main difficulty with the knife-against-roll type of coating machine is a tendency for small particles to be caught under the knife and cause scratches and breaks in the web. These particles may come from imperfectly filtered lacquer or from the surface of the paper. This condition may be minimized by proper filtration of the lacquer and by brushing the paper before it reaches the knife. Another disadvantage is that the lacquer spills over the surface of the supporting roll when a break in the web occurs. The machine must then be stopped and the roll thoroughly cleaned before it can be started again. If any spots remain on the roll, they will show as lightly coated areas on the paper each time the roll turns one revolution.

The surface of the roll must be kept in good condition and precautions against chipping or deep scratching should be taken because these imperfections may show as photographic images in the finished coating, especially on very thin paper.

Coater with Knife Against Plate

A further modification of the knife coater is the introduction of a finely ground and polished metal plate directly under the face of the knife. This feature has the same function as a supporting roll. In this type also, the uniformity of coating depends on the accuracy of the surface of the plate and that of the face of the knife. As the plate is rigid and the paper must be pulled over it, this arrangement has more tendency to cause breaks, especially in thin papers. Its only advantages over the supporting roll are lower cost and easier replacement and maintenance. In general, the supporting roll is to be preferred.

Coater with Knife Against Rubber Apron

The supporting roll or plate is often replaced by a driven rubber apron that passes under the coating knife. Because the rubber apron is elastic, it has less tendency to trap foreign particles underneath. It is also better adapted for paper that has slack and uneven edges. In case of a break in the web, it presents the same cleaning problem as a supporting roll. A natural rubber apron becomes softened, after some use, by the solvents in the coating. A synthetic rubber apron is better in this respect.

Roll Coaters

Many variations of roll coater exist, but only a few will be discussed here (see Figure 5-3). Fundamentally, in this type of coater, the solution is applied to the paper by a roll. In its simplest form it consists of a roll that revolves in the lacquer, which is transferred to the paper by direct contact. In its more complicated forms, other rolls, which act as doctor rolls, metering rolls, transfer rolls, and backing rolls are added. Some of the common types are described briefly as follows.

Pressure Coater

This very simplest form of roll coater consists of two rolls. Roll A revolves in the lacquer and deposits it by direct contact with the paper. Roll B acts as a backing roll. Both rolls can be made of metal, or roll B can be made of rubber.

Fig. 5-3. Reverse-roll coating line (*Courtesy of Waldron-Hartig Division, Midland-Ross Corporation*).

Kiss Coater

This is almost like the pressure coater except that roll A deposits the solution on roll B, which in turn deposits it on the surface of the paper. Roll C acts as a backing roll and is usually made of rubber. The amount of coating deposited depends on the pressure between rolls A and B and also on the viscosity of the lacquer used. The uniformity of the coating depends on the accuracy of the rolls and on the uniformity of the paper thickness.

Reverse-Roll Coater

Probably the most versatile type of roll coater is the reverse-roll type, Figure 5-3. One roll, the doctor roll, meters the solution to the coating roll, which applies it to the paper supported by a rubber backing roll. The amount of coating deposited is determined by the pressure of the doctor roll against the coating roll; the pressure can be adjusted by micrometer screws. The characteristic feature of this type of roll coater is that the coating roll is driven in a direction opposite to that of the paper.

One undesirable property of roll coaters, especially the direct-roll type, is that they have a tendency to deposit the coating with ribs or striations. The larger the depositing roll, the coarser these striations are likely to be. The size of the striations is also a function of the viscosity and surface tension of the coating solution. This effect is largely eliminated in the reverse-roll coater by the reverse action of the coating roll.

Spray Coaters

Recently, some successful applications of very thin coatings on rough paper have been made by spraying methods. In general, however, spray coating is seldom used and is unsuitable for most paper-coating applications. The economy is poor because of the necessity for using slower-evaporating solvents, and relatively low solid concentrations.

Air-Brush Coaters

The air brush consists essentially of a coating roll, which picks up the solution from a bath and deposits it on the paper, and a compressed-air nozzle extending over the full width, which blows off the excess lacquer with a controlled stream of air. The air brush is ideally suited for aqueous coatings, but it is not widely used for solvent coatings. The general opinion in the solvent coating industry is that the blast of air is likely to cause skin drying, roughness, and bubbles in the coating because of the rapid evaporation of the solvent. It also has a tendency to blow off particles of coating, which become lodged in the nozzle. However, one coating manufacturer, who uses the air brush chiefly for aqueous coatings, claims highly successful applications with solvent solutions.

Print Coaters

Many variations of print coaters are built on the principle that an engraved or etched roll picks up the solution and transfers it either directly to the paper or to a rubber offset roll, which in turn deposits it on the paper. The amount of coating deposited depends on the design of the engraving and its depth. Print coaters are suitable only for light coatings with low-viscosity solutions. They can apply uniform coatings, but the quantity of solution applied can be varied only within narrow limits. Because the lacquer is deposited in a discontinuous film, it must retain sufficient fluidity, after being applied to the paper, to flow out before it becomes permanently set by the evaporation of the solvent. Print coaters are usually operated at higher speeds than either roll coaters or knife coaters.

General Comparison of Knife Coaters with Roll Coaters

Knife coaters can apply solutions of much higher viscosities than roll coaters. Reverse roll coaters probably operate best in the viscosity range 300—1,500 cP. Knife coaters can handle solutions of 10,000 cP with no particular difficulty and can be cleaned more quickly and easily between color changes. Roll coaters, however, have a lesser tendency to cause scratches and breaks, because they do not have the sharp constrictions present in knife-type coaters. Roll coaters usually operate at higher speeds. The best types of either class can apply coatings of good uniformity that can be adjusted with close precision.

MANUFACTURING HAZARDS

In coating with solvent solutions, two hazards are encountered that do not exist when aqueous solutions are used. These are the danger of fire and explosion from the use of inflammable solvents and nitrocellulose, and the possibilities of toxic reactions on the human body from exposure to excessively-high solvent concentrations. However, with an understanding of the problems involved and with the proper precautions, these hazards can be reduced to a minimum. Some coating plants have run for years without serious fires or explosions and without adverse effects on the employees. It must be emphasized, however, that these hazards always exist and slackness in precautionary measures must never be allowed.

Fire and Explosive Hazards in Drying Solvent Coatings

All the commonly-used solvents, with the exception of some chlorinated compounds, are highly inflammable. In connection with inflammability and explosion, three terms that have great significance are commonly used. These are flash point, fire point, and explosive limits.

Flash Point

The flash point of a solvent or of a mixture of solvents is the lowest temperature at which enough vapors are given off to form a flammable mixture of vapors and air immediately above the liquid surface. In a mixture of solvents, this temperature may be higher or lower than the flash point of the lowest-flashing constituent. Its value is largely a function of the vapor pressure characteristics of the mixture involved. Flash points of some of the common solvents are given in Table 5-59.

Fire Point

The fire point is the lowest temperature at which the solvents or lacquer give off enough vapors to continue to burn after being ignited. It is usually somewhat higher than the flash point and is determined by several factors; the heat of combustion, the supply of oxygen, the rate of heat diffusion, and the type of surface presented.

Explosive Limits

The explosive limits of a solvent or solvent mixture are the minimum and maximum concentrations of its vapor in air at which ignition can take place on the application of a spark or flame of sufficient intensity and become self propagating. The minimum concentration is called the lower explosive limit; the maximum, the upper explosive limit. Below the lower limit ignition will not take place, even though a spark or flame is present, because there is insufficient vapor. Above the upper limit ignition will not take place, because the mixture is too rich and contains insufficient oxygen. The lower explosive limit is the critical concentration that must be considered in solvent coating.

One must be able to calculate the lower explosive limit of a mixture and be able to make measurements of vapor concentration in the drier and in the recirculating and exhaust ducts connected with the drier.

White[25] found the following explosive limits in percent by volume for the solvents listed.

Thornton gives the following explosive limits, in percent by volume, for aliphatic hydrocarbons, which are found in petroleum diluents.

Durrans[27] found that the temperature affects the lower and upper explosive limits, and cites the following results for a toluol-air mixture:

TABLE 5-61

Explosive Limits of Common Solvents

	Lower (20°C)	Upper	
Ethyl Ether	1.71	48.0	(20°C)
Acetone	2.89	12.95	(20°C)
Methyl Ethyl Ketone	1.97	10.2	(20°C)
Benzol	1.41	7.45	(60°C)
Toluol	1.27	6.75	(60°C)
Methyl Alcohol	7.05	36.5	(60°C)
Ethyl Alcohol	3.56	18.0	(60°C)
Ethyl Acetate	2.26	11.4	(60°C)

Test conditions have an effect on the results obtained and the size

TABLE 5-62[26]

Explosive Limits of Common Hydrocarbon Solvents

	Lower	Upper
Pentane	1.35	4.5
Hexane	1.10	4.2
Heptane	0.95	3.6
Octane	0.84	3.2
Nonane	0.74	2.9
Decane	0.67	2.6

TABLE 5-63

Effect of Temperature on the Explosive Limits of a Toluene— Air Mixture

Temperature	Lower limit	Upper limit
50°C	1.5	4.85
300°C	1.3	6.20

and shape of the container, direction of flame propagation, source of ignition, humidity, turbulence, temperature, and pressure are all important factors.[28]

The lower explosive limit (L.E.L.) of a mixture of solvents may be calculated from the formula of Le Chatelier* as follows.

Given n_1, n_2, n_3 . . . , respectively, as the mol fraction of each solvent in the mixture, and N_1, N_2, N_3 . . . , respectively, as the lower explosive limit of each solvent, expressed as a percentage, the lower explosive limit of the mixture is given by the formula

$$\text{L. E. L.} = \frac{1}{\dfrac{n_1}{N_1} + \dfrac{n_2}{N_2} + \dfrac{n_3}{N_3}}$$

In practice, sufficient air must be exhausted from the drying system to maintain the vapor concentration at a safe fraction of the lower explosive limit. Practice varies from plant to plant, but a reasonable limit is 20—25% of the lower explosive limit. Some coating plants, at times, operate at concentrations considerably above this figure to reduce the heat losses from the drying system by exhaust. In any case, the concentration should not be allowed to go above 50% of the lower explosive limit. A continuous vapor-sampling meter, which gives an alarm when the concentration rises above a set limit, should be used. With a portable explosion meter one should also probe various points in the drier and in the ducts, when localized pockets of high concentration are suspected. These instruments are made by at least two reli-

* Coward and Jones of the Bureau of Mines state in Bulletin No. 279 that in general this formula is very nearly correct, but that there are some marked exceptions and it must not be applied indiscriminately.

able manufacturers in the United States.

As an example of the method of calculating the lower explosive limit and the amount of air to be exhausted, the following problem is presented:

A nitrocellulose solution containing 30% solids is to be coated on 48 in. wide paper. The solvent mixture contains the following solvents:

	% by volume
Petroleum Naphtha	30
Ethyl Alcohol	25
Ethyl Acetate	30
Acetone	15

	Lower explosive limit %	Molecular weight	Weight/gal, lb
Naphtha	1.15	110	6.20
Acetone	2.89	58.1	6.60
Ethyl Alcohol	3.56	46.1	6.57
Ethyl Acetate	2.26	88.1	7.50

Basis—100 gal solvent

	gal		lb		Mole	Mole fraction
Naphtha	30.0	=	186	=	1.69	0.177
Acetone	15.0	=	99.1	=	1.705	0.179
Ethyl Alcohol	25.0	=	164.3	=	3.565	0.375
Ethyl Acetate	30.0	=	225.0	=	2.554	0.268
			674.4		9.514	

The speed of the paper is 40 yd/min. The amount of coating to be applied is 35 lb of wet lacquer per ream of 500 sheets 20 in. $\times$ 26 in. How much air must be exhausted to keep the vapor concentration at 20% of the lower explosive limit? Let us assume that the vapor is uniformly mixed with the air.

$$\text{Lower explosive limit} = \frac{1}{\dfrac{0.177}{1.15} + \dfrac{0.179}{2.89} + \dfrac{0.375}{3.56} + \dfrac{0.268}{2.26}} = 2.27\%$$

$$\text{Average molecular weight} = \frac{614}{9.51} = 70.9$$

$$\text{Solvent evaporated per minute} = \frac{40 \times 36 \times 48 \times 35 \times 0.70}{500 \times 20 \times 26} = 6.51 \text{ lb}$$

$$\text{Moles evaporated per minute} = \frac{6.51}{70.9} = 0.0918$$

$$\text{Solvent vapor evaporated per minute (Measured at 0°C and 760 mm)} = 0.0918 \times 359 = 33.0 \text{ ft}^3$$

At 20% of the lower explosive limit of this mixture, the percentage concentration in the air $=0.2 \times 2.27 = 0.454\%$. The volume of vapor-laden air which it is necessary to exhaust

$$= \frac{100}{0.454} \times 33.0 = 7270 \text{ ft}^3/\text{min (at 0°C and 760 mm).}$$

$$= 7270 \times \frac{460+70}{460+32} = 7830 \text{ ft}^3/\text{min (at 0C° and 760 mm).}$$

How much air would have to be exhausted if a concentration of 50% of the lower explosive limit were maintained in the drying system?

$$\text{Volume necessary} = \frac{0.2}{0.5} 7830 = 3130 \text{ ft}^3/\text{min (at 70°F and 760 mm).}$$

These calculations were based on pure compounds. Commercially available solvents are of course not pure, but the error introduced by this assumption is not great.

General Safety Measures Against Fire and Explosion

The purpose of all safety measures is twofold: to prevent sparks or flames in any area where there are inflammable materials, and to prevent the accumulation of any inflammable solvents, solids, or vapors in any area, except where they must be kept and used. Sparks may arise from static electricity, from the friction of steel against steel, or from improper electrical installation. Flames may arise from spontaneous decomposition, spontaneous combustion, localized over heating, or from violation of the rule against smoking and carrying matches in a coating plant.

Electrical Equipment and Wiring

In areas where solvents or nitrocellulose are stored and used, all electrical work should adhere strictly to the Fire Underwriters Code for Class I, Group D installations. The following general rules should be observed:

All wiring should be in conduit, with explosionproof fittings, which are properly sealed at appropriate points.

All motors, starting boxes, and motor controls should be of Class I, Group D, explosionproof construction.

Lights should be explosionproof and not merely vaporproof.

All other electrical equipment, such as bells, alarms, temperature controls, and tachometers, should be explosionproof.

Storage

Large quantities of solvents can be kept more safely and economically in storage tanks if facilities are available for obtaining delivery in tank trucks. Local and state ordinances and fire laws pertaining to storage vary and should always be consulted. Tanks may be buried and covered with 2 ft of dirt, or they may be set on concrete piers above ground. Each method has its advantages and disadvantages. Buried tanks are safer with regard to fire, but leaks are difficult to detect, and repairs are impossible to make unless the tank is uncovered. Elevated tanks are exposed to fire hazards, but maintenance and leak detection are much simpler.

If tanks are buried where ground water is likely to seep in, they should be securely anchored by guy wires attached to heavy blocks of concrete to prevent the possibility of floating to the surface. The dirt underneath the tank should be carefully sifted and freed from sharp stones, that might cause rupture. Buried tanks should be placed far enough away from any building, to prevent any possible leakage from seeping underneath the building and forming dangerous pockets.

Elevated tanks should be surrounded by dikes of more than sufficient capacity to hold the solvent contents in case of leaks. They should be grounded by means of heavy insulated wire attached to copper-clad ground rods, at least 10 ft long and $\frac{3}{4}$ in. in diameter. Each tank also should be equipped with a supplementary copper tiller rope connection, 10—15 ft long, which can be placed in contact with the truck unloading solvent into the tank.

The inlet pipe to the tank should go almost to the bottom to prevent the formation of static electricity, which is likely to occur in a freely falling stream of solvent. The suction pipe should be at least 2 in. from the bottom to prevent rust and water from entering into the line. The vent pipe should be provided with a gauge hatch and flash arrester.

In transferring solvents from storage tanks to the operating building, it is safer practice to pump from the tank to the building, which is at a higher elevation, than to flow the solvent by gravity into the building.

If solvents are stored in drums, they should, if possible, be kept out of the direct rays of the sun and away from areas where they will be exposed to passing vehicles and pedestrians. Before being brought into the mixing room, the drums should be swept carefully to reduce the possibility of fine pebbles and sand falling into the mixing tanks and creating sparks.

Nitrocellulose drums should be stored either in the shade or under a roof, to prevent the development of excessive temperatures inside the drums in the summer.

Handling and Mixing

Pumps used for moving solvents from storage tanks should be non-sparking in construction and should be driven by explosion-proof motors. If solvents are handled in drums, nonsparking tools should be used. Nonsparking forks and wrenches should always be used for handling drums of nitrocellulose.

Mixing tanks should be provided with solvent inlets that lead through a pipe to within a few inches of the bottom. If this is impractical, the solvent should not be allowed to fall freely to the bottom, but should be introduced in such a way as to hit the sides of the tank to minimize the formation of static. Floor areas around the mixing tank should be nonsparking and conducting. As pointed out already, one should operate the mixing tank with a self-contained, direct-driven mixing unit, in a completely enclosed, explosion-proof and sparkproof housing. If a belt drive is used, precautions can be taken against static formation by treating the belts with a conducting carbon dispersion and by removing any static with carbon brushes. The tank should also be provided with an adequate vent, preferably fitted with a flash arrester. An automatic carbon dioxide system that can discharge into the top of the tank in case of fire provides good insurance. All mixing tanks, of course should be properly grounded.

Common practice calls for mixing batches of clear lacquer in large mixing tanks and pumping the clear solution into small mixing containers for pigmenting and further treatment. Portable, fractional-horsepower, propeller-type agitators are often used for these smaller batches. These should always be explosion-proof and have heavy insulated connecting cords that plug into explosion-proof receptacles. Portable agitators with flexible cords are never as safe as fixed agitators with all electrical connections in conduit. However, they are so useful and easy to handle that this sacrifice to safety may be justified if proper precautions are taken to insure that the connections are always in good condition.

Coating

All tools around the coating machine should be nonsparking and the floor areas near the coating head should be nonsparking and conducting. A minimum amount of the coating solution should be exposed. If the coating solution is pumped to the machine, or flows by gravity from a large reservoir, an automatic shut-off should be provided for the lacquer in case of fire. The coating machine should be adequately grounded and provision should be made to remove static by the use of conducting chains or static eliminators. An automatic carbon dioxide system should be provided at the coating head.

Static Eliminators

Static eliminators are of two general types—electrical and radioactive. In the electrical type, positive and negative ions are formed. These bombard the surface of the paper and neutralize the static electricity. The radioactive types give off rays, some of which neutralize the static. One type of radioactive static eliminator consists of a ribbon of gold foil in which radium salts have been imbedded. α, β, and γ rays and radon gas are given off. The radon gas is sealed in the foil and connot escape. The α rays neutralize the static electricity and have an effective range of about 3 in. The β and γ rays have no effect on the static electricity, but are continually discharged and form a health hazard that must be guarded against. With proper controls this type of eliminator appears to be safe to use and the Occupational Hygiene Departments of the states of Massachusetts and New Jersey have approved their use when proper safeguards are provided. These departments should be consulted before installations are made. A new type of radioactive eliminator, in which polonium is used, has recently appeared. It has the advantage of giving off only α rays, but it must be renewed after a year.

Drying

The problem of explosion hazard in the drying chamber has been mentioned. Explosion doors or vents should be provided in the ratio of at least 1 ft^2 to every 15 ft^3 of content of the drier. Driving chains should be on the outside of the chamber to eliminate the possibility of sparks. Lights inside the drier should, of course, be explosion-proof.

If a hot-air system is used, the fan should be provided with a non-sparking impeller. The exhaust fan should be provided with a device that will automatically shut down the coating machine if the fan stops through motor or mechanical failure.

General Protection against Fire

Fire extinguishers and fire hoses should be sufficient in number and conveniently placed and every employee should be instructed as to their location and use. An adequate sprinkler system must be provided to meet insurance requirements.

Toxicity Hazards

It is undoubtedly better for a human being to inhale fresh air than air containing solvent vapors of any kind. Organic solvents, however, probably will always be used in one way or another and even with the best ventilation some vapors will be present in the air in solvent-coating plants. The user should have a reasonable knowledge of the physiological effects of the solvents used, of the allowable concentrations for safe operation, and of the proper means of ventilation to keep the concentration as low as possible.

Some solvents have acute narcotic effects that pass away in a short time after exposure. Other solvents have cumulative toxic effects. The physiological effects depend on several factors: the nature of the solvent, its concentration, the length of exposure, and the individual exposed. Some persons became immune to the effects of specific solvents after prolonged exposure; however, this is not a safe assumption to make in any case.

The problem of determining the maximum allowable concentrations of solvents is a difficult one. It involves the study of exposure of human beings over a long period of time, and it is understandable that some of these data may be conflicting and questionable. The Massachusetts Division of Occupational Hygiene has made a critical examination of the available information and gives maximum allowable concentrations (in parts per million by volume) for various organic solvents (Table 5-64).

TABLE 5-64

Maximum Allowable Concentration of Solvent in Air from Toxicity Standpoint

Benzol	35
Toluol	200
Xylol	200
Methanol	200
Petroleum Naphtha	1000
Butyl Acetate	200
Amyl Acetate	200
Carbon Tetrachloride	100
Ethylene Dichloride	100

Although the author of the report gives no figures for acetone, ethyl acetate, and alcohol, he states that he would class them with naphtha.

Besides the chlorinated solvents, which are not used frequently in paper coatings, the most toxic solvent is probably benzol. It is best to avoid its use if possible. If this is not feasible, adequate ventilation must be provided to keep its concentration, in areas of exposure, below 35 parts per million, and periodic physical examinations should be made of employees who work with it.

In its report "Chronic Exposure to Benzol," the Massachusetts Division of Occupational Hygiene published studies of scores of cases of benzol exposure and cited as its chief toxic effect damage to the blood and bone marrow, with eventual death in some cases.

With regard to a comparison between benzol and toluol, Elkins states, "We are frequently asked how these two compare in toxicity. There are distinctly two schools of thought on this subject. One holds that they are equally injurious, but that because of its lower vapor pressure, the concentrations of toluol that exist in the air are much lower than those of benzol, and for this reason there are few cases of toluol poisoning. The other school claims that toluol lacks entirely the chronic action that makes benzol so deadly and that as long as concentrations that might cause acute poisoning are avoided contact will not be injurious. We have taken a middle course."[28]

With regard to methanol, the US Bureau of Mines has reported that occasional exposure is not dangerous, but that continuous exposure to high concentrations is. The commonly used ester solvents (such as ethyl acetate and butyl acetate), ketones (such as acetone and methyl ethyl ketone), and petroleum naphthas apparently have no cumulative effects, but their concentrations should be kept below the figures stated to avoid acute narcotic effects. Dioxane and mesityl oxide, which are used to some extent as vinyl resin solvents, may produce adverse physiological effects and should be used with care.

Two problems arise in attempting to keep solvent concentrations well below the allowable maximum: (1) the proper method of ventilation and (2) measurement of the concentration.

The best method of ventilation is to remove the vapors in the areas where they originate and where they are the most concentrated. In the mixing room, exhaust hoods or ducts should be placed as closely as possible to any exposed mixer. If solvents are used for washing small mixing containers, the washing operation should be isolated and the solvent vapors removed at the source. At the coating machine, sufficient make-up air should sweep by the operator into the drier (to compensate for the exhaust) to remove solvent vapors coming from the bank or bath of lacquer at the coating head. General ventilation is not nearly as effective as ventilation at the source, and it is often useless unless the fan is close to the floor level. Exhaust fans near the ceiling are seldom effective and they can even be harmful, since they may sweep solvent-laden air past the breathing zone of the workmen.[29]

Measurement of concentration can be made by at least two methods. In one method, a definite amount of air is drawn through a calibrated flow meter and passed through a weighed or standardized tube containing an appropriate chemical solution for absorbing the specific vapor. In the second method, use is made of an interferometer. Its operation is based on the index of refraction of gases and is described in the Bureau of Mines Technical Paper No. 185.

REFERERENCES

1. Hercules Powder Co., "Nitrocellulose" (1944).
2. Tennessee Eastman Corp., "Eastman Cellulose Esters" (1945).
3. Hercules Powder Co., "Cellulose Acetate" (1945).
4. Hercules Powder Co., "Ethyl Cellulose" (1944).
5. Dow Chemical Co., "Ethocel Handbook" (1946).
6. Bakelite Corp., "Vinylite Polyvinyl Acetate Resins" (1942).
7. Bakelite Corp., "Vinylite Copolymer Resins for Surface Coatings" (1942).
8. B. F. Goodrich Chemical Co., "Geon Polyvinyl Materials" (1947).
9. Dow Chemical Co., "Saran F-120 Resin" (1947).
10. Bakelite Corp., "Polyethylene Resins" (1948).
11. Hercules Powder Co., "Parlon" (1946).
12. H. R. Thies, "Pliolite—A Coating Material for Paper," *Paper Trade J.* **108** (8), 96-102 (1939).
13. Corn Products Sales Co., "Zein" (1946).
14. Plasticizer Chart, *Modern Plastics Catalogue*, Plastics Catalogue Corp., New York (1947).
15. "Plasticizers," Circular No. 485, Scientific Section of National Paint, Varnish, and Lacquer Association, Washington, D.C.
16. Simonds and Ellis, *Handbook of Plastics*, Van Nostrand, New York (1943) p. 251.
17. W. M. Munzinger, *Technologie der Weichmachungsmittel*, J. F. Lehmans, Munich Verlag (1934).
18. M. C. Reed, "Behavior of Plasticizers in Vinyl Resins," *Ind. Eng. Chemistry* **35**, 896 (1943).
19. J. J. Mattiello, *Protective and Decorative Coatings*, Vol. II, Wiley, New York (1944).
20. H. A. Gardener and G. G. Sward, *Physical Examination of Paints, Varnishes, and Lacquers*, 10th ed., Henry A. Gardner Labs. Bethesda, Md. (1946).
21. C. P. Bogin, in *Protective and Decorative Coatings*, J. J. Mattiello, Vol. I, Wiley, New York (1941) p. 644.
22. White, *J. Chem. Soc.* 1462 (1919); 1244 (1922).
23. J. R. Thornton, *Phil. Mag.* **33**, 190 (1917).
24. T. H. Durrans, *Solvents*, Van Nostrand, New York (1931).
25. H. B. Elkins, "Toxic Fumes," Mass. Div. of Occupational Hygiene, Boston, Mass.
26. W. A. Gumaer, "Ventilation of Heavier than Air Vapors," Barrett Products

Div., Allied Research Products, Inc., Baltimore, Md.
27. Bakelite Corp., "Vinylite Resins," (1944).
28. Hercules Powder Co., "Benzyl Cellulose" (1944).
29. Lewis, Squires, and Broughton, *Industrial Chemistry of Colloidal and Amorphous Materials*, Macmillan, New York (1942).
30. J. J. Mattiello, *Protective and Decorative Coatings*, Vol. I-V, Wiley, New York (1941–1946).
31. Packaging Catalogue Corp., *Packaging Encyclopedia*, 1946–47 (1948).
32. Plastics Catalogue Corp., *Modern Plastics Encyclopedia* (1947).
33. F. A. Simonds and H. Ellis, *Handbook of Plastics*, Van Nostrand, New York (1943).
34. F. Zimmer, *Nitrocellulose Ester Lacquers*, Van Nostrand, New York (1934).
35. Bogin, Kelley, and Maroney, "Prevention of Gelling of Bronze Lacquers," *Ind. Eng. Chem.* **23**, 982 (1931).

chapter 6

Hot-Melt Coatings

BERT C. MILLER AND PAUL H. YODER

During the early stages in the development of plastic resin coatings, the production of a paper base material with either a high utility or a high decorative value necessitated the use of lacquers or solvent coatings. Such coatings were made with a volatile solvent that requires drying after application, with the resultant loss of the solvent unless a recovery system is used. Two well-known methods of coating papers (other than with resins) by means of a high-temperature technique, were generally employed, i.e., waxing and asphalting, but neither yielded coatings as satisfactory as a resinous film.

INTRODUCTION

A hot-melt coating is one in which all of the solid material that might be in a solvent-base formulation is used, but is made sufficiently fluid for coating machine application by means of heat rather than a solvent. However, this procedure is not quite so simple as the definition makes it appear. Many of the materials that are used in solvent coatings do not become fluid on heating, others char and burn, and some are not miscible without the use of solvents. Hot-melt technology as a method of coating paper is entirely different from all others. Application is similar to that of asphalting and waxing; the results obtained are those of solvent-coated papers but hot-melting is a distinct technique and does not generally replace any of the others.

Hot melts are composed of various resins and resinous materials (both natural and synthetic), crystalline and amorphous waxes, cellulose derivatives, and plasticizers. Occasionally included are some of the proteins, which are miscible when heated to form blends that liquefy and have the necessary viscosity to be applied at specified tempera-

289

tures. After the coating is laid down on the paper surface and chilled, it immediately becomes dry and hard and the composition deposited on the base is in the form of the parent material.

The materials must be nonpolymerizing and nonvolatile, must not change color with heat, their viscosity must not change under prolonged heating, and above all they must be chemically stable over extended periods at temperatures above that of application. Some coatings are applied at 400-450°F, but the use of such hot melts is limited, as the base paper is usually harmed to some extent by this extreme heat. Melts with an application temperature above 300°F are in the high-melting-point class; those in the range 200-300°F are in the medium class; and those in the waxing range (150-210°F) are in the low-melting-point class.

Several pertinent fundamental factors should be considered from the standpoint of economics. Melt formulations contain no volatiles; no solvents and no vehicles of any kind are required for application. All the coating applied to the paper remains there. Because no drying is required, the coating operation is finished immediately after application, allowing time for the paper to reach a normal temperature. The hot-melt method demands comparatively little machine space and the speed of the machine is not dependent on the drying of the web. These facts favor compactness and high operating speed, depending on the flow of heat and the degree of chilling required by the particular product.

Certain disadvantages in the use of hot melts must be borne in mind. Relatively high temperatures are required; therefore the piping, heating system, and the machine itself must be well insulated and guarded to prevent burns. Temperatures must be closely controlled; a certain tolerance exists, but with more than the allowable variation, the viscosity of the material may change and cause either too much or too little coating to be applied. Hot melts are compounded for application at specified temperatures; the viscosity and the machine should be set for these temperatures, within the allowable tolerances.

This problem is not difficult if the proper controls are used on the equipment, regardless of whether steam, electricity, or circulating oil supplies the heat. Further, when hot-melt coating is done at high temperatures on a roller coating machine, the applied film may show local distortions because of variations in temperature. A slight distortion of the roll sharply alters the weight of the applied coating laid down at almost 100% solids. On the equalizer type of machine, this condition is minimized, but variations in the tension of the paper across the web may also adversely affect the weight of coating.

The time required for the setting of the melt and the degree of chilling are largely dependent on the speed of the machine and the melting point of the coating. With low melting point materials, which usually have a high wax content, chilling is necessary to obtain a glossy surface finish. For high melting point materials, which are largely formulated with resins, chilling may not be necessary to obtain a high gloss. Chilling the coating and the paper is essential to prevent the sheet from being rewound hot and causing blocking in the roll. Chilling is extremely important: The higher the speed of the coating machine and the higher the input of heat into the coated sheet, the more chilling capacity must be available to remove this heat and to return the sheet to room temperature.

HOT-MELT FORMULATIONS

Blends that have low, medium, and high melting points are considered.

Low Melting Point Blends (150-210°F)

Blends with low melting points are formulated largely for use on a conventional waxing machine where the application temperature can be higher than that required for waxing. These materials are generally waxy in nature and the added properties improve the coating in several ways.

Increasing the gloss and hardness of the paraffin coating and improving scuff resistance may be desirable, to give the wax an effect similar to that of a transparent film. Hence one adds resins that do not increase the viscosity greatly and that can be applied at a temperature only slightly higher than that required for the straight wax. Compatible resins that improve the scuff resistance of paraffin wax are: Modified rosins, coumarone–indene, modified polystyrene resins, and low-molecular-weight polyethylene, because they generally have low melting points and do not appreciably increase the viscosity of the blend.

The addition of materials that increase the tackiness and adhesive power of the wax coating readily impacts the properties of sealing by means of a paper-tearing bond. These additives may be heat-sealing-type resins, such as polyamides, cyclized rubbers, isobutylene polymers, butyl rubber, and ethylene–vinyl acetate resins. Pale crepe rubber can also be used, but preferably it should be milled into the wax. These additions should also reduce strike through of the coating during the heat-sealing operation. Natural or synthetic rubber gives increased film flexibility and prevents the ordinary paraffin coating from cracking when applied heavily.

Many substances can be used to prevent the usual "blushing and blooming," often encountered on waxed papers. The various glycerides can be substituted for a portion of the wax. These additions increase the hardness of the film; they are perfectly compatible with paraffin wax and absorb a portion of it, thus reducing the blushing. Polyethylene is also used to harden the film, to increase the resistance to blocking, and when properly chilled, to improve the gloss. Stearates also reduce the blush and bloom, but must be used in limited quantity if one is to avoid a greasy surface.

Medium Melting Point Blends (200-300°F)

The majority of the second classification of hot-melt compositions is primarily aimed at approaching more nearly the characteristics of plastic resin coatings. The wax, or soft element, in these formulations is reduced to a minimum, and the required fluidity to apply the resins, is obtained by heating up to 300°F. This class of material is usually waterproof, flexible and block resistant, and can be employed within a viscosity range that permits the application of light and heavy coatings. The surface of the film, if properly applied and chilled, has high gloss and luster. As these materials are generally resinous, the particular requirements over and above the basic characteristics are obtained by a variation of the base resins or combinations of them. For example, one may desire a sheet that has all the usual characteristics plus a high degree of resistance to grease; therefore, grease-resistant resins and plasticizers are included in the formulation.

Many resins effect a strong heat seal; heat-sealing resins are incorporated in the formulation so that the coating is heat-activated and forms a seal within a narrow temperature range. If the difference between the application and sealing temperatures is slight, the sheet can sometimes be dusted with starch or talc to prevent blocking, but these are usually unnecessary.

Resistance to the transmission of water vapor is an important trade requirement for certain functional packaging papers, and various agents can be incorporated for obtaining this characteristic. If the basic characteristics that hot melts give to the sheet are carefully considered, any other particular specification can easily be met by proper compounding and formulation.

High Melting Point Blends (300-450°F)

High-temperature blends are generally made without the use of paraffin or other low melting point ingredients. They are formulated with film-forming resins and plasticizers. The resulting product is a completely self-supporting film that has practically no penetration into the base other than that required for adhesion. The film can be extremely hard and scuff resistant. It can be made as flexible and noncracking as the base to which it is applied. These coatings are clear, transparent, and are considered competitive with the better-grade solvent coatings and film laminations.

Although many standard hot-melt compositions are available, one usually desires a formulation compounded to suit a particular requirement. The film-forming materials include the higher-melting-point synthetic and cellulose-base resins because they impart good flexibility and eliminate any tendency to block. Film-forming materials that require the highest temperature for efficient operation can be used in melts run at lower temperatures by proper formulation with waxes and resins. As a result, the favorable properties of the film-forming material are only slightly reduced, but the costs and running viscosity drop considerably. Therefore, a melt formulated only from a film-forming material and plasticizer should not be used if this is not an absolute requirement. The compounded melts are in general more common than are the straight resin-plasticizer formulations.

Some of the specific materials used in hot-melt formulations are discussed here, but detailed formulas for specific applications are not included because they vary extensively, depending on their applications.

Resins

Polystyrenes are colorless and their properties make them valuable in the formulation of many different types of coatings and plastic compositions. They are stable to heat and light and can be readily compounded with vinyl resins, cellulose esters and ethers, ester gums, hydrogenated or modified rosin, pure phenolics, alkyds, and maleic esters.

Vinyl acetate resins are used extensively in hot melts and provide certain highly desirable characteristics. They are relatively stable to both heat and light and are resistant to normal aging. They are light in color and are suitable for use in heat-sealing compounds as they provide excellent adhesion and a tough film. When compounded with other types of resin, the film is resistant to alkalis, acids, water, and grease and may have a high luster.

Polyethylene, readily compounded with other resins and waxes, is water-white and forms a clear, transparent film. If polyethylene dominates in the formulation, a strong film results. It is relatively grease resistant, has excellent resistance to water, good resistance to water vapor, and excellent flexibility at low temperatures. It is compatible with all hydrocarbon waxes and a wide range of natural and synthetic resins. In comparison with many other film-forming materials the cost is low; therefore it has many applications in coating paper and paperboard.

The polyamide resins are used extensively for heat-sealing, particularly if heat sealing at low temperatures is desired. These resins are resistant to water, strong alkalis, mild acids, oils, and greases. They are comparatively dark and if a light-colored coating is desired, they must be formulated with other resins of a lighter color. Castor oil is a valuable plasticizer for these resins but it is rather incompatible with paraffin wax. When a small amount of paraffin wax is added, resistance to water vapor is increased and the blocking point is elevated. Rosin esters are suggested for compounding with the polyamides.

Cellulose acetate butyrate is valuable as it is relatively noninflammable. It has good resistance to heat, light, and moisture, and is extremely hard and resistant to abrasion. Its odor, somewhat objectionable, can occasionally be masked.

Ethyl cellulose has been employed extensively because of its compatibility with so many resins and waxes and its film-forming and heat-sealing properties. When compounds made with ethyl cellulose are used, the operating temperature of the machine must be carefully controlled. If overheated, ethyl cellulose becomes dark and the viscosity of the melt decreases rapidly. Because of its comparatively high cost, it is being replaced by lower-cost film-forming materials.

Chlorinated diphenyls are extremely valuable in the formulation of hot-melt coatings because of their wide range from soft to hard and their compatibility with many film-forming materials. They are light in color, nonoxidizing, permanently thermoplastic, and noncorrosive. They are resistant to burning and have excellent electrical properties.

The use of a high polymer made up of amino acids plays an important part in the compositions of hot melts for some particularly exacting uses. When compounded with hydrogenated rosins, alcohol-soluble ester gums, and alcohol-soluble phenolics, the resulting film is desirable for printing-ink applications. This film is highly resistant to abrasion, has excellent adhesion, and lends itself to formulation for cold-set printing inks. It tends to char and discolor unless it is used at carefully controlled temperatures. The specific temperature depends on the nature of the compatible resins and plasticizers.

Vinyl butyral resins are light in color and have strong thermoplastic properties. When properly compounded and chilled they have excellent gloss characteristics. Suitable modifying agents are castor oil glycerides and plasticizers of the phosphate and phthalate types. The comparatively high cost of these resins preclude their use for many applications.

Ethylene–vinyl acetate copolymer resins are light in color and have a wide range of compatibility with petroleum waxes, rosin-derived resins, butylastic polymers, terpene-modified phenolic resins, and modified styrene resins. Because of their compatibility with waxes, coatings have low resistance to water vapor at low coating thicknesses and at consequent low cost.

Acrylic resins are water white, but generally make better adhesives than coatings because of their softness and tackiness. By proper formulation they can be compounded into excellent heat-seal coatings. They are resistant to alkalis and mild acids, are tough, and are good electrical insulators.

Many resins are used in formulating hot melts simply to reduce the cost. These resins should be light in color, sparkling clear, ranging from "N" to "WW" on the US Rosin Color Standard. They must hold the color on prolonged heating and be compatible with other resins and waxes. Modified maleic resins, ester gums, and many others belong to this group.

Plasticizers

Plasticizers must be selected for the desired coating result and for compatibility with a specific resin. Some plasticizers are based on low-melting-point resins or rosin oils. Synthetic plasticizers are made by many chemical companies that can usually recommend a plasticizer suitable for the particular resin to be used. Some of the better-known plasticizers are dibutyl phthalate, dibutyl tartrate, and tricresyl phosphate.

Polybutene or polyisobutylene of high molecular weight, or butyl rubber, may be used in the formulation of hot melts to lend particular advantages. They are compatible with rubber, paraffin waxes, ester gums, and some of the natural gums. They may be milled into other ingredients on a rubber-compounding mill. They retard oxidation or hardening with age; when formulated with wax or ester gum, they make a good adhesive. If used in sufficient quantity in the formula, pressure-sensitive coatings can be obtained.

Low-molecular-weight polyethylene and polypropylene are good plasticizers for many hot-melt formulations.

Pigments

Pigments in hot-melt coatings must be ground into an ingredient of the formula, usually the plasticizer, because this operation is simpler than milling them into the resin itself. The grind should be as concentrated as possible so that the formula is not overbalanced with plasticizer. The pigments can also be dispersed directly into the resin in a Banbury mixer, for example.

Titanium dioxide is the best white pigment because it provides great opacity per unit weight. An opaque white film can be produced with 18-20% of titanium dioxide by weight.

Fairly large quantities of colored pigments that withstand application temperatures are used in both coatings and inks, but their cost is considered high when they must compete with a previously colored sheet and a clear coating. Some recommended pigments are iron blue, phthalocyanine green and blue, chrome green and yellow, cadmium red, and zinc yellow.

Waxes

Waxes with a large range of melting points can be used in hot-melt formulations. For example paraffin wax of 133-135°F melting point is satisfactory for the lower-melting compounds, whereas for higher-melting compositions, waxes with melting points as high as 200 °F may be desirable.

The amount of wax needed depends entirely on the end use of the product and on what is desirable from an application and cost standpoint. Wax reduces the viscosity of the coating and acts as a lubricant for the resin component during the compounding operation.

Amorphous waxes are employed to some extent, again depending on the end use, but these have a tendency to soften the film and do not chill with a high gloss. When amorphous wax is used, its melting point should be in the range 170-185°F.

Carnauba and candelilla waxes often add hardness to the film and serve to even out the coating. They do not scratch as easily as the usual waxes.

Waxes cannot appear in any composition that must be completely scratchproof or have good printing characteristics.

COATING EQUIPMENT

Machinery used for applying hot-melt coatings depends largely on the type of material to be applied and the surface to be coated. Although all types of hot melts will run on specially-built machinery special machinery is not necessary for some of them.

As melts can be classified into high-, medium-, and low melting-point materials, one must consider what standard equipment can be used for melts that fall in the various temperature ranges.

The average wax coating machines have an application range of 180-190°F, but often the range can be stepped up to 200-225°F, where many of the usual hot melts can be satisfactorily handled. In general, at these temperatures, the speed of operation must be less than that of the normal waxing operation. The smoothing bar is not essential for many materials that will run on a waxing machine. These melts should be chilled and the conventional waxing machine should be used. If two-side coating is desired, the web is reversed and passed over a second application roll similar to the first.

For the harder, more resinous materials, operating temperatures must be above those permissible for the waxing machine. Special types of equipment, to be discussed later in some detail, must be designed for these materials.

Gravure Coaters

Gravure coaters operate on the principle of an etched or knurled roll that runs in the melt bath and picks up an excess of the melt. The excess is removed from the surface of the roll by an oscillating doctor blade or scraper, leaving the melt, which is to be deposited on the paper, in the cells of the roll. The paper web is then brought into contact with the roll by means of a rubber impression roll and the coating material deposited on the paper surface in the form of dots corresponding to the cells in the coating roll. The coated side of the sheet must then be passed over a smoothing bar, which reheats the tops of the dots and spreads them out into a continuous film. This operation usually requires a higher temperature than that at the point of application. The method is desirable for high-speed continuous operation on one stock or for one large run at a constant coating load.

The coating weight can only be changed by altering the pattern on the coating roll. The widths of the web and the rubber impression roll must always be equal. This kind of machine is usually heated with circulating oil and should be capable of temperatures as high as 450°F to accomodate all types of melts.

"Equalizer" Coater

The most popular type of hot-melt coating machine in general use is probably the "equalizer," on which a single application roller runs in the coating bath. The paper passes over this roller and picks up an excess of coating. The excess is removed by a rod that rotates in the same direction as the paper web, but at approximately one-quarter of its speed. The rod is wire-wound, with the gauge of the wire determining the amount of coating left on the paper. The tension of the web across the machine should not vary. The coating must then be smoothed to eliminate the wire marks and to polish the surface. Smoothing and polishing should be done by means of a smoothing bar operated at a slightly higher temperature than that of the applicator roll. The smoothing bar does not reduce the weight of coating and should not be confused with the doctor or scraper.

Roller Coater

The roller coater is another type of applicator machine. The web to be coated passes between two heated rolls that are adjusted to meet the requirements of the coating weight and the thickness of the stock. In common with other methods, the material should be smoothed after application to eliminate the roller marks. These machines can be operated at higher speeds than those of the "equalizer" type. Care must be taken in the operation of these units because the rolls, running together at high temperature, have a tendency to distort and to run hotter at the ends than in the center. Because of the resulting difference in expansion, the rolls may become smaller in diameter in the middle and thus lay down more material on the center of the sheet than on the edges. The best solution is to have the rolls ground in at the desired running temperature.

Sheet Coating

Hot melt materials can be applied to flat sheets, although this technique is not generally of much interest to the converting industry. Roller coaters are ordinarily employed for this purpose, but because the sheet cannot be held under tension as on a web coating machine the smoothing section must be of different construction. The operation is accomplished by a driven polished roller that runs in the reverse direction to that of the sheet and by a pulling roller that rides on the back of the sheet to make the contact. These machines are particularly designed for the better grade film-forming materials where speed is not an important consideration.

Carton blanks can be coated with hot melts that have a wide range of viscosities. Modified wax coatings of low viscosity are applied to blanks in a manner similar to that for application of straight paraffin. Then the blanks are passed through a cold water bath to set the coating and develop the highest gloss. High-viscosity hot melts are applied by means of pattern waxers that provide either full coverage or coverage only on exposed areas, leaving glue flaps uncoated. To develop the highest gloss, coated blanks can be run through a remelter or against a polished chrome-plated surface, followed by chilling.

Hot melts can also be applied to sheets of paper or paper board by silk screen application so that the coating is applied only to the desired areas.

Fig. 6-1. Hot melt, or curtain, coater [1] (*Courtesy of Vibro Mfg. Co.*).

Fig. 6-2. Hot melt, or curtain, coater [2] (*Courtesy of Vibro Mfg. Co.*).

By means of curtain coating, paper and corrugated board with a caliper of more than 0.010 in. can be coated with hot melts of a wide range of viscosity. The hot melt is forced either by gravity or by pressure through a slit of predetermined width to form a curtain that falls on the sheet as it passes underneath. Such coatings are relatively thick because of the necessity of maintaining the curtain. See Figures 6-1 and 6-2 for hot melt or curtain coaters.

Cast Coating

Hot melts are also applied by casting. Instead of applying the melt directly to the paper stock, it is first applied in a fluid condition to a continuous, polished surface—either a drum or an endless metal belt. The belt is then gradually chilled as it moves forward; while the material is still in a plastic stage, the paper sheet or web is brought into contact with the coated face of the belt. When the chilling is complete, the sheet of paper and the cast film are stripped as one sheet from the polished surface. The advantages of this type of coating, which is in reality a lamination of the film, are the resulting high gloss and smoothness of the film because the surface is a mirror reproduction of the very smooth polish of the belt. This method also eliminates all penetration except that needed for bonding, allowing the use of porous papers as well as fabrics and permitting even the bridging of openings such as those in window cartons.

Dip Coating

The dip method is employed for coating papers on both sides or when an actual impregnation of the sheet with the melt is desired. A machine for this type of work is relatively simple. The web is led into a bath of molten material and is then drawn up vertically, passing through wire-wound equalizers to doctor and level the coating on both sides of the sheet. The proper tension is obtained by the pressure of the web on the two levelling bars of the equalizers. The sheet is passed through smoothing bars in a similar arrangement and immediately chilled.

This machine offers the further advantage that two other dry webs can be brought into contact with the two coated faces while the melt is still in the fluid stage and under a conventional pressure roller, so that a three-ply laminated sheet can be produced.

HEATING EQUIPMENT

The usual method of obtaining the necessary high temperatures is circulation of hot oil or a substitute. This is not a difficult system of heating; the major problems are good circulation with no dead spots and no leakage. A pump must be employed to circulate 4—15 gallons of heating fluid per minute, depending on the speed of the machine and the amount of heat-transfer medium to be heated. The application tank, premelt tank (if present), the application roll or rollers, the equalizer bar, and the smoothing bar should all be connected so that the oil can circulate evenly through all pipes, with suitable valves to impart different temperatures where necessary; for example, the smoothing bar should be slightly warmer than the application roller. Rod-type electrical heating units are sometimes used for spot heating the smoothing bar or equalizer rod. The method of heating the oil depends on the installation; electricicty is the simplest means and is the most easily controlled. An oil burner may be less expensive to operate and presents no problem if it can be conveniently placed near the machine operation.

Steam is usually employed for heating the lower-melting-point materials. Steam at pressures of 30—80 lb/in.^{2} is sufficient for treating the lower-melting-point materials and some formulations in the middle temperature range. Electrically-operated strip heaters can also be used but they must be placed to prevent localized overheating and subsequent burning of the coating.

COATING BASE PAPERS

One of the outstanding characteristics of hot-melt coating technique is that it usually is a purely surface application and the base paper need not necessarily be of the finished types. However, the better- and smoother-surfaced papers produce the best results, since the less the penetration of the coating beyond that required for bonding, the smoother the coating because of the better continuity of the film. Coating compounds of lower melting points are more fluid and have a tendency to penetrate the base to a greater degree than the more viscous or high-melting-point materials.

The most desirable base paper is a sheet with a high surface density and finish. The overall density of the paper need not be high, but a specific surface density can be attained by either supercalendering or by the application of a pigment coating. For example, glassine paper lends itself well to all types of hot-melt coatings as the surface is extremely impervious and allows the coating material to form a film readily on the surface. A machine-glazed paper that may appear to have a fine surface does not possess a high surface density; hence the lower-melting point, low-viscosity coating materials readily penetrate into it and more coating must be applied to obtain the desired end result.

With proper operation of coating equipment for hot melts, viscosity can be controlled by means of the temperature, so that porous papers can be satisfactorily coated by applying the coating in a viscous, semiplastic stage.

Hot-melt coating compounds rarely affect the printing on the surface of the paper because they do not contain water or other solvents. Practically any type of printed paper can be coated without bleeding or smudging of the print. The main problem is discoloration that is due either to penetration or excessively high temperatures.

The previous statements refer to conventional coating methods. When cast coating is employed, the base paper is of practically no consequence as the hot-melt coating is completely preformed and applied in a plastic state to any base, whether porous or not. Cast coating can be used with any base flexible enough to pass through the coating rollers; the coating is always a preformed film surface.

APPLICATIONS

Hot melts have excellent possibilities for decorative coatings. The coatings are clear and transparent and one of their outstanding characteristics is high gloss. Gloss depends on the base to which the coating is applied and on the thickness of the coating, but compared with other techniques for application of plastic coatings finishes with higher gloss can be obtained with lower coating loads. Other properties that can also be built into such coatings are flexibility, resistance to discoloring by light, hardness, scuff resistance, and resistance to oxidation on ageing. Inasmuch as hot melts produce a strictly surface coating with little penetration, porous papers or papers that are soft in texture lend themselves well to coating for decorative use.

The wide application of hot melts is due to their functional properties, such as heat sealability and resistance to water vapor, water, grease, weak acid and weak alkali. Because 100% solid materials are used, the base in not damaged by the drying out of water or solvents after the film is deposited. No bubbles, blisters, pinholes, or penetration into the paper should occur if the application is properly made. In compounding the coating formulation for a high degree of functional characteristics, hot melts offer one of the best methods of application for transfering the coating to the base sheet. Large quantities of kraft paper, bleached carton board, and corrugated board are manufactured by this process for the packaging industry.

Where high resistance to both grease and water vapor is essential, it is sometimes feasible to laminate a highly grease-resistant sheet such as glassine and then to coat either or both sides of the lamination. Fundamentally, waxes or waxlike materials are useful to furnish resistance to water vapor, however, they are also soluble in oils, fats, and greases. This difficulty can be overcome to some extent in the compounding, but if an extremely high degree of resistance is required, a double coat or a coated and laminated structure is generally provided.

The double-coat method is sometimes employed where extreme flexibility is desired. The prime coat saturates and softens the base paper; the second coat supplies protective characteristics. A reasonable assumption is that with proper compounding and proper application, hot-melt coatings can produce resistance equivalent to that of other flexible coatings, films, and foils.

One must not assume from the separate treatment that the functional and decorative values of hot-melt coatings cannot be combined. A coating with high gloss and attractive surface appearance can also be highly resistant to grease and water vapor. Most of the resin coatings have good functional characteristics in themselves, which are to a great extent due to the unique method of application.

BIBLIOGRAPHY

William H. Neuss, ed. "Testing of Adhesives," *Tappi* Monograph Series, No. 26 (1963).

chapter 7

Paste Dispersion Coatings

W. D. HEDGES

Vinyl dispersion resins have become an important factor in the vinyl resin and plastics field since their introduction during the early part of 1940 to the protective coating industry in the United States.

VERSATILITY

The versatility of these dispersion resins was first established in the cloth-coating industry, and millions of yards of cloth coated with vinyl dispersion are now produced for the manufacture of upholstery, for use in the interior trim of automobiles, for wall coverings, and for garment material. Items made by slush and rotational moulding techniques have produced a large market for these resins. They have found their way into production of both supported and unsupported vinyl foam products. The rapidly growing use of these resin compounds can be attributed to their unique properties, which permit economical manufacturing techniques not available in other forms to the vinyl industry.

In general, vinyl coatings applied by paste dispersion techniques offer the same advantages that vinyls offer over other surface materials. Often, special properties, that cannot be incorporated by any other processing technique can be imparted by means of paste dispersion.

ORGANOSOLS

When the technology of applying vinyl resins from paste-like coatings was first introduced, there was a distinct difference in the techniques required for the handling of organosols and plastisols. One of the first resins suitable for this technique could only be suspended in a mixture resulting in approximately 60% nonvolatiles. Those familiar with vinyl compounds realize that if so much liquid were required to produce 100% solids consisting of plasticizer only, the film would be entirely too soft and tacky for any practical application. Therefore, to develop a film of real value, part of the plasticizer had to be replaced by a volatile solvent. This type of paste dispersion was given the name of organosol.

An organosol results when insufficient plasticizer is present to form a paste and fluidity in the system must be attained by the addition of a volatile diluent.

PLASTISOLS

Later as a result of new and improved polymerization techniques, resins were developed with a particle size distribution that would permit the production of a fluid coating mixture at practically 100% solids by reducing the amount of liquid required. These resins enabled coating mixtures to be prepared from plasticizers only and at a sufficiently low level to produce a satisfactory coating. The term plastisols was given to this type of dispersion coating.

PROCESSING TECHNIQUES

Since the early days of this development, many improvements have been made in polymerization technique whereby greater control of the particle size, shape, and distribution by the resin manufacturers have enabled them to supply more uniform material to the formulators. Many new plasticizers have been made available, thus offering greater versatility in formulation. New techniques for dispersing the resin in the liquid phase, coupled with a variety of surface-treatment agents, have contributed to the improvements. All these modifications have tended to reduce the original wide differences in processing technique required for preparation of an organosol and plastisol. The difference is so small in most paste dispersion coatings today that it is hard to define where an organosol stops and a plastisol takes over.

Control of Viscosity

Because these coatings are not true solutions, they do not follow the Newtonian law and their viscosities must be evaluated by rheological techniques. One of the first methods for developing a proper paste dispersion mixture was by controlling its viscosity at a given shear. In the past this was chiefly done by controlling the amount of volatile liquid. With the advent of the improved resins and the other improved features mentioned previously, viscosity now is commonly controlled not so much by the addition of nonvolatile material, but by selection of a resin to give the desired property. Varying the type of plasticizer with addition of surface active agents also further refines and controls the properties of the coating mixture.

Even a change in the type of pigment or filler in a given formulation can so affect the viscosity that small amounts of organic solvent must be added to obtain the desired flow characteristics and thus convert what is normally a plastisol to an organosol. One seldom needs to add more than 5 or 6% of organic solvent to develop paste dispersion coatings that cover a wide range of rheological properties.

Control of Agglomerate Size

Techniques for processing organosols and plastisols varied considerably. Normally reduction of the agglomerate size of the resins in the organosol was attained by lengthy grinding in a ball mill. Where sufficiently high-boiling, nonsolvent extenders were used, roller mill techniques were employed. Even during the early stages of this development, reduction of the size of the resin particles was often necessary when the plastisol approach was used. Here, because of the absence of volatile solvents, roller techniques were usually employed for reducing agglomerate size.

Organosols and plastisols are so closely related to each other that in further discussion they can be treated as one and will be called paste dispersion coatings.

VINYL DISPERSION PASTE COATINGS

The use of vinyl dispersions applied by coating techniques offers a distinct economic advantage because of the production of a coatable mixture at high solids content. One can also deliver considerably more weight in one application than can be supplied by the same vinyl from a solvent system.

Advantages

Vinyl-paste dispersion coatings have made other important contributions to the converting industry. When one must add solvent to make the coating more workable, the solvents required are less expensive than those normally used for vinyl solution coatings and make considerable savings possible for those convertors that must evaporate expensive vinyl solvents without the benefit of a solvent-recovery system. In addition, for those that must make multiple passes of the sheet to build up sufficient coating thickness, a method is available by which vinyl coatings 1/8-in. or greater in thickness can be applied in one operation.

Properties

Even these advantages might be considered by some to be over-shadowed by the unusual properties offered by these coatings. They are resistant to mineral acids (including aqua regia), organic acids, and alkalis. They also resist soap and washing. They are unaffected by grease, butter, cheese, oil, and many chemicals, both wet and dry. They can be made flame-, light-, and weather resistant. They are clear, neutral, nonoxidizing, nonyellowing and nonheat-reacting, non-toxic, odorless, and tasteless. They are insoluble in many solvents, including alcohols, and therefore resistant to liquors and wines; they are also resistant to petroleum hydrocarbons. They have excellent resistance to water and through recent technological advances they can be made almost permanently flexible. Finally, they can be produced in all colors from soft violets to brilliant reds.

Formulation

Unlike vinyl solution coatings, vinyl-paste dispersion coatings are formed by a technique in which the resin is dispersed or suspended as small distinct particles in an organic medium that offers limited solubility at room temperatures to the vinyl resin particles. These vinyl resin particles form a stable and easy-to-handle fluid coating mixture.

The introduction of this concept represented a radical departure from the previously accepted technique of applying vinyl coatings from a solvent mixture. The formulation and application of aqueous dispersions or latices parallels that of vinyl dispersions; this parallel can be used to introduce the technology of vinyl-paste dispersion. A brief note on latices is given first.

LATICES

Latices are of many kinds: natural rubber latex, synthetic rubber latex, and vinyl latex. All latices have one property in common: they are a suspension or dispersion of small discrete particles in water. These latex particles are generally spherical with varying degrees of hardness, depending on the latex involved. On evapora-tion of the water, the particles coalesce with or without additional heat to form a continuous film.

Vinyl paste dispersions are roughly similar to latices, i. e. vinyl paste dispersions are suspensions or colloidal dispersions of indivi-dual discrete resin particles in a liquid medium. However, the liquid medium for vinyl pastes is not water, as is the case with latex, but an organic liquid. This organic liquid medium serves to peptize and to suspend permanently the dispersed resin particles. The organic medium is actually too poor a solvent at room temperature to dissolve the dispersed resin particles, but is capable of penetrating the resin particle surface sufficiently to hold it in permanent suspension. The organic medium may be volatile, such as a solvent, or nonvolatile, such as a plasticizer. The dispersed vinyl resin particles are generally assumed to be spherical, or nearly so, and offer a large surface area in relation to their weight.

Because of peptization, they do not stick together or agglomerate, but remain suspended as discrete and individual particles approximately $0.01-1.0\,\mu$ in diameter, floating around in the organic medium. Upon the application of heat either from evaporation of volatile solvents or from the increased solubility of the resin in the organic liquid, the resin particles fuse together to form a tough and continuous film. After fusion, they cannot be redispersed.

TECHNOLOGY OF PASTE DISPERSIONS

When a soluble organic resin is dissolved in a suitable solvent, the mechanics of solution follow the pattern of wetting, swelling, and finally dissolving. If the solvent has some affinity for the resin, but is not sufficiently polar (does not readily dissolve the resin), a gelled mass is the end result of attempted solution. This is practically the situation in the formation of the dispersed particles in a paste dispersion. The dry resin before dispersion consists of hard, discrete particles, less than $0.5\,\mu$ in diameter, that have become agglomerated into masses of approximately $50-100\,\mu$ in diameter. During dispersion in an organic liquid, these masses are deagglomerated, and the individual, discrete particles are suspended in the liquid.

Swelling

The particle size after dispersion depends on the polarity of the liquid system used to disperse the resin. The greater the tendency of the resin to soften in the organic medium, the larger the particles will be. The degree to which the surface of the resin particles of a paste dispersion is softened by the medium is one of the most important characteristics of a paste dispersion. The degree of swelling is further controlled by the type of plasticizer and plasticizer-solvent combination used in the formulation. Improper softening of the surface of the paste-dispersion resin particles is believed to be responsible for many of the difficulties encountered in securing films free from pinholes, cracks, crazing, and blisters.

Components

The simplest paste dispersion consists of two components: (1) the paste dispersion resin and (2) the organic liquid. The dispersing organic medium is usually a plasticizer or combination of plasticizers. These are capable of peptizing the vinyl paste dispersion resins; in this instance, the plasticizer is called the dispersion agent. It is also possible to use an active, organic solvent as a dispersing medium and to utilize a nonsolvent or poor solvent for the paste dispersion resin; in this case the nonsolvent is called the diluent.

The combinations outlined here must be carefully selected and used in definite proportions depending on their chemical nature. Any deviation from the thoughtfully chosen proportions usually results in a poor paste dispersion system. The physical properties of the fused film are determined by its solid content and are independent of the volatile solvents in the formulation as long as these are correctly balanced.

Dispersing Agents

Both the dispersing agents and diluents affect the degree of swelling. The dispersing agents have the stronger peptizing effect because of their greater tendency to soften the paste dispersion resin. By proper selection of dispersing agents and diluents, the desired results can be achieved. In the formulation of vinyl paste dispersions, there is usually at least one dispersing agent and one diluent (volatile or non-volatile).

Some of the most commonly used dispersing agents for vinyl paste-dispersion coatings are dioctyl phthalate, di-2-ethyl hexyl phthalate; other members of the phthalate family; tricresyl phosphate and other phosphates; and, if active organic solvents are desired, such solvents as isophorone, cyclohexanone, methyl ethyl ketone, octyl acetate, diisobutyl ketone, and trimethyl nonanone. Some diluents in order of their decreasing swelling effect are toluol, xylol, higher-boiling aromatics, and the aliphatic hydrocarbons. Those aliphatic hydro-carbons that are higher in naphthenic compounds make better diluents than those that are essentially paraffinic in nature.

Any dispersing agent or mixture of dispersing agents can be used with any diluent or mixture of diluents. Therefore, if one wishes to produce a paste dispersion with excessively softened resin particles, he can usually select a suitable organic liquid as the dispersing agent.

Plasticizers

If a 100% solids formulation were required, a particular plasticizer would be needed. If this gave a viscosity above the applicable range, one could reduce the viscosity by introducing suitable surface-active agents to the system without deviating from the 100% solids formulation.

Conversely, a system with only slightly swelled resin particles would result from the use of organic liquid of poor swelling value and an aliphatic hydrocarbon. By choosing a plasticizer rather than an active organic solvent as the organic dispersing medium, the formulator has a double-edged tool in that a plasticizer acts as a dispersing agent and also as a softener, with the plasticizer becoming a permanent part of the vinyl film. As far as the degree of swelling is concerned, plasticizers would fall approximately in the middle of the best and poorest swellers.

If a limitation on the amount of plasticizer in a given formulation is necessary, the formulator has two choices to obtain a workable system: (1) to supplement the liquid portion with a solvent dispersion agent or (2) to use a surface-active agent to produce a formulation within workable limits.

The choice of the proper plasticizer or plasticizers for softening vinyl resins is an art in itself. Space does not permit a discussion of all the possible ramifications. The important characteristics of a satisfactory plasticizer include fastness to light and heat, clarity, nonspewing on aging, nonmigrating to other finishes, permanency, and absence of odor and toxicity. The ability to flexibilize a vinyl film at low temperatures is often desired. These properties are usually found in mixtures of plasticizers.

Any plasticizer or mixture of plasticizers compatible with vinyl dispersion resins can be employed. One of the most commonly used is di-2-ethylhexyl phthalate, but almost any of the phthalate plasticizers can be incorporated if it gives the desired properties. Tricresyl phosphate, trioctyl phosphate, and other phosphate plasticizers are satisfactory if lack of light resistance is not important. Other plasticizers such as adipates, azelates, and certain polymers are used when low temperature flexibility and a degree of permanence are required.

Plasticizers are incorporated into formulations to impart flame resistance and to give flexibility to the final coating, while not exuding to the surface upon extended aging periods under both natural and artificial conditions at both low and elevated temperatures. Each plasticizer contributes certain properties and the choice of plasticizers is largely dependent on the specific problem.

The simplest form of paste dispersion consists of two components —one of which is the organic liquid. If the organic liquid is an active organic solvent, the use of the correct solvent composition cannot be over emphasized.

Paste Dispersion Viscosity

Figure 7-1 indicates the dependence of the paste dispersion viscosity on solvent composition. The logarithm of the viscosity is plotted on the vertical axis; the solvent composition or indirectly the softening tendency of the organic solvent composition is plotted on the horizontal axis. Thus softening tendency of the organic solvent mixture for the resin increases from left to right.

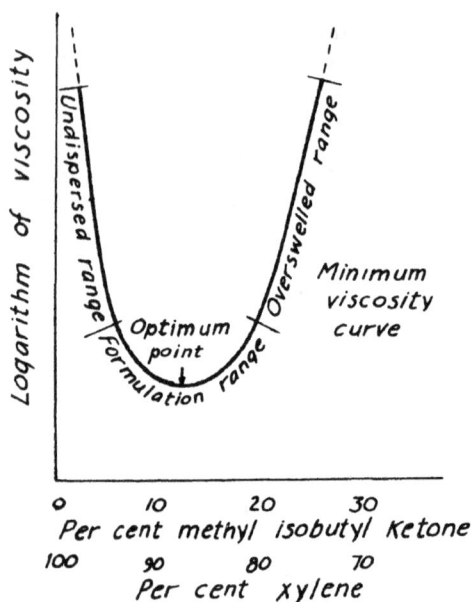

Fig. 7-1. Dispersion minimum viscosity curve for methyl isobutyl ketone and xylene.

In this instance, methyl isobutyl ketone was selected as the prime solvent and xylol as the diluent. If such a diagram is required, it can be prepared by making a series of dispersion formulations at the same concentration of solids and progressively varying the solvent content.

Figure 7-1 indicates that as the percentage of active solvent is increased, the viscosity, which in the first instance is very high, decreases rather quickly to the minimum point when the solvent combination is 10% methyl isobutyl ketone and 90% xylol.

A change in the solvent composition to give an increased amount of active solvent and a lesser amount of diluent results in an increase in viscosity. All this, it is emphasized, occurs without a change in dispersion resin solids.

Formulations that fall in optimum regions of the curve give the most desirable features with respect to stability, flow, and good coating characteristics.

If the dispersant component of the simple paste dispersion is a plasticizer, the same fundamental approach is used, i. e. the softening tendency of the dispersion system is varied until the optimum point is reached.

RHEOLOGY

See Chapter 3, "Rheology of Paper Coatings and Instruments for the Measurement of Their Flow Properties."

MANUFACTURE OF PASTE DISPERSIONS

After the paste dispersion formulation has been properly selected in the laboratory, the next step is to translate this laboratory information to a production scale. Formerly, such an operation might have tended to be a frustrating and expensive experience, but new methods both in formulation and laboratory equipment have minimized this problem so that today laboratory batches can be correlated with production batches of the same material. Even with modern technology, modifications or adjustments to production formulations must be made on the machine to make them suitable for application.

In the production of paste dispersion formulations problems may occur that result in the coatings being (1) too grainy or containing particles brought about by poor dispersion, (2) too high or too low in viscosity, or (3) unstable during storage, which would produce coatings too viscous for correct application to active plasticizers.

Corrective measures would have to be taken to ensure that such material could be used in production, and examples of steps that would be necessary are as follows:

1. To correct a coating that is too grainy or contains particles incorrectly dispersed, grind on a three-roll mill if the viscosity of the resulting formulation lends itself to such a method of grinding; if it

does not, such grinding methods as ball milling are acceptable, or passing the faulty material through a high-speed rotor-stator type of equipment, always ensuring that the setting between rotor and stator is sufficiently wide so that gelling does not take place during the additional grinding requirements.

2. To correct too high viscosity, add diluents or combinations of diluents and active solvent mediums or surface active agents. The choice of additive is governed by the formulator and would be based on judgment in light of the end use, final solids, and final viscosity requirements. To correct too low viscosity, add small quantities of a highly absorbent inorganic compound. Here again the amount is governed by end viscosity and subsequent coating properties.

3. To correct instability during storage that results from the use of too active a plasticizer or other organic medium, conduct further laboratory studies relating to storage features and reformulation of the coating.

EQUIPMENT

Paste dispersion coatings can be manufactured by several methods. In the early days of this type of coating formulation, the paste dispersion resin, organic liquid medium, pigments, extenders and surface-active agents were mixed thoroughly in some type of mixing equipment capable of handling the product.

Pony Mixer

For formulations of high viscosity a pony, dough mixer, or similar device with a relatively low agitator speed is used. Resin, organic dispersing agent, pigments, fillers or extenders, and surface-active agents if required are added to a suitable container and thoroughly mixed at low speed.

Three-Roll Mill

After thorough blending, the whole formulation is run through a three-roll mill. In general the more passes made through a three-roll mill, the more satisfactory the resulting product; at least two passes are usually necesssry to obtain a satisfactory coating medium.

Pebble, Stone, and Ball Mills

Where the dispersing agent was an active organic solvent, with or without a diluent, the older method of manufacture was by utilizing pebble, flint, or steel ball mills. The capacities of such rotating mills varied from one to 20 or 25 drums. The pebbles were usually made of porcelain; the flint stones, imported Baltic flint; and the steel balls, of a special alloy. The linings in the pebble or stone mills were generally of porcelain, but burrstone lining was sometimes used. Mills with burrstone linings were difficult to clean, particularly when changing colors.

Each of these mills has its particular advantages. The porcelain balls and lining are desirable because they are easy to clean, but the porcelain tends to abrade slowly and to contaminate the organosol, thus causing hazy films. Flint stones in a porcelain-lined mill have the same disadvantage. Flint stones in a burrstone-lined mill produce coatings of excellent clarity, but have the previously mentioned cleaning defficulties. Steel balls, because of their greater density, produce exceptionally well-ground vinyl dispersions and clear films, but the vinyl dispersion is apt to be contaminated by iron salts, which cause discoloration when the dispersion is subjected to heat or ultraviolet light. Fortunately, iron-contaminated vinyl dispersions can be stabilized to make them acceptable for certain uses. Any of these mills, if properly used, produces satisfactory vinyl dispersions.

High-Shear Equipment

As more and more technology is applied to this field, newer and improved equipment is made available for paste dispersion coatings. Of great help is the improved technology in the manufacture of the paste dispersion resins themselves and now one can make a coating suitable for production purposes simply by stir-in techniques. With the improved technology in resin manufacture, the use of high-shear or high-impingement equipment has become an accepted method of obtaining a satisfactory paste dispersion coating.

Grinding-In Pigments

To incorporate colored pigments into the basic formulations may or may not be necessary. If they are required, they can be added in one of two ways: (1) careful weighing of the colored pigments into the initial formulation and (2) dispersion of color concentrates of pigments in a plasticizer acceptable for use in the end paste dispersion followed by addition of the pigment plasticizer grind to the basic formulation.

If rigid color controls are required, the most generally accepted method of arriving at batch color control is to make (1) a base paste-dispersion coating, usually uncolored or white, normally having a lower plasticizer content than that required in the finished formulation, and

(2) a concentrated pigment paste. The pigment color pastes are made by grinding the required pigment in the appropriate plasticizers. To produce a paste dispersion coating of a certain color, the base paste dispersion is tinted with the pigment color paste. At the point when the correct color is achieved, adjustments to the resin/plasticizer ratio may or may not be necessary.

When passing a paste dispersion coating through a three-roll grinder becomes necessary, the formulation should have sufficient cohesion and adhesion so that it will pass through the mill by sticking to the rolls.

If high-speed equipment is chosen as the means of manufacturing paste dispersion coatings, the total formulation may or may not need to be cooled during mixing procedures so as to decrease the tendency of the total formulation to gel partially.

Cycling the paste mixture is sometimes practiced when flow characteristics must fall in a narrow range of temperature. In this way one can obtain the correct cohesive properties at a predetermined viscosity.

MECHANISM OF FILM FORMATION

The successful coating of paste dispersions is largely dependent on a thorough understanding of the mechanism by which the individual resin particles, dispersed in a suitable organic media, are converted into a continuous film. The vinyl resin particles are in a dispersed state, i. e. they are surrounded by an organic liquid as a dispersion medium. This organic liquid can be a plasticizer, a plasticizer plus a suitable organic diluent, or a plasticizer plus an active organic solvent as the dispersing medium.

Plasticizer as Dispersion Medium

If the organic medium around the resin particles is a plasticizer, therefore giving a coating composition of 100% nonvolatile material, the formation of a coating with optimum physical properties is relatively simple. The coating need only attain a temperature at which the resin particles are solvated by the plasticizer so as to fuse together to form a continuous film. The correct selection of resin and plasticizer at levels relative to each other largely governs the conditions necessary to attain a continuous film and also to attain a coating that is neither too soft, too hard, or too tacky to meet the specific end requirements. Modern techniques provide a relatively easy way to obtain coatings that have a wide range of plasticizer content and that can be applied to a variety of substrates.

Plasticizer and Diluent as Dispersion Medium

If the coating contains both plasticizer and diluent as the organic medium, film formation involves (1) the evaporation of the diluent and (2) the fusing together of the resin particles. Only solvated resin particles can be fused into a continuous film. If at any time before the paste-dispersion resin particles are fused they are permitted to return to their original hard and unswollen state, they cannot then be fused into a continuous film. Fortunately most paste dispersion coatings contain sufficient amounts of plasticizers to keep the resin particles in a peptized state even after most of the diluent has been evaporated.

Plasticizer and Organic Solvent as Dispersion Medium

If the coating consists of a suitable organic solvent plus a plasticizer, the three basic mechanisms of fusion are: (1) by the action of a plasticizer, (2) by the combined action of the plasticizer and a dispersing agent, and (3) by the action of the dispersing agent alone. The fusing mechanism is dependent on the type of coating medium. The first type of fusion is associated with coatings that have a relatively high content of plasticizer. The second type of fusion is concerned with coatings that fall in the medium plasticizer range; the third type is exemplified by coatings that contain considerable organic solvent but little or no plasticizer.

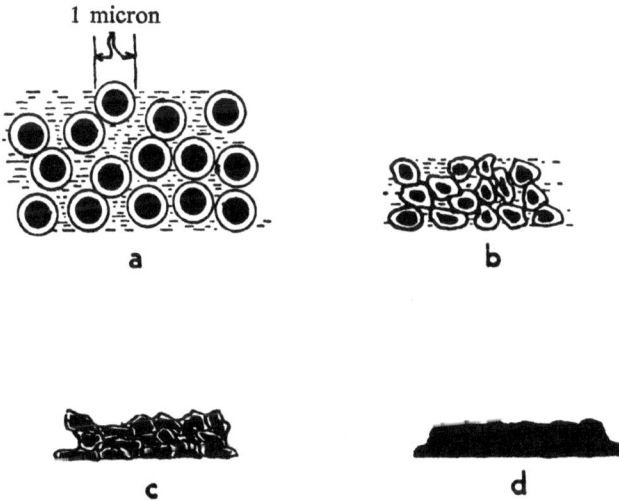

Fig. 7-2. Organosol fusion.

In every case, the resin particles must be heated to a point where these particles can flow together and yield a continuous solid film. In Figure 7-2 the resin particles are in their peptized condition surrounded by the organic medium. In *b* the heat is causing the particles to become more fluid and some of the dispersing solvent is being lost. In *c* the particles have almost flowed together and in *d* the film has been formed.

Some solvation of the resin particles in the organic medium occurs even at room temperature; the heat applied to the coating medium further softens the particles, which then become plastic and flow together.

Fusion by Plasticizer

With the first type of fusion mechanism, where there is sufficient amount of plasticizer to cause the resin particles to become fluid during the heating process, the coating is generally fused by first evaporating or releasing most of the volatiles at some low temperature and then fusing the resulting coating at a higher temperature. The temperatures required are dependent on the volatiles being used and also on the relative molecular weight of the resin. Zoned ovens are ideally suited for this type of operation where the first zone is maintained at a relatively low temperature and the second zone or fusing zone is maintained at a higher temperature.

This operation can be carried out in separate units the first being the solvent release oven where the volatiles are evaporated; the partially fused coating is then wound into a roll which is taken to the higher temperature oven for more complete fusion of the coating.

Fusion by Plasticizer and Dispersing Agent

The second type of fusion mechanism, where both plasticizer and dispersing agent are necessary for fusion because of relatively low plasticizer content, requires a different type of oven technique. The resin particles must be fused together before any appreciable amount of organic solvent, which acts as a dispersing agent, has had an opportunity to evaporate. To accomplish this, the resin particles, the plasticizer, and the solvent must be heated to a temperature at which the particles become fluid.

During this heating process the volatile portion is evaporating. If all the volatile portion is allowed to evaporate before the fluid temperature is reached, the particles cannot become fluid enough to fuse and poor film formation results. The recommended procedure is then to place the paste dispersion coating into an oven at a relatively high temperature. By doing so, the coating is quickly brought to the fusion point before the volatile solvent has had an opportunity to escape completely.

Because some plasticizer is in the formulation to aid the fusion process, some range in drying temperatures exists. The usual procedure is for the coating to enter the oven at 325—350°F and bake for 45—60 sec. Any additional baking at this temperature is superficial since the solvent is usually evaporated by this time. After the initial baking cycle, the paste dispersion coating is exposed to a high temperature for a relatively short time to complete the fusion. Exposure to relatively high temperatures for a short time does not necessarily correct bad techinques in film fusion when the pasted dispersion coating is of this type.

Fusion by Dispersing Agent

The third type of fusion is dependent on the dispersing agents alone. This type of fusion and coating is relatively rare, but it will be briefly discussed. The fusion of a coating with a low plasticizer content might seem to be extremely difficult. As formulations of this type do contain high-boiling organic solvents, they do allow sufficient opportunity for the resin particles to reach the fusion point before the solvents evaporate. Normally this type of fusion requires a quick, high-temperature bake. The unfused coating should be run into an oven at 360—380°F and held there until all the solvent is evaporated. Because the solvents have high boiling points, the danger of bubbling or pitting of the surface is minimized.

The temperatures mentioned previously are normal for achieving a completely fused state. The ideal temperatures for each formulation are governed largely by the individual ingredients of a given formulation, but the actual fusion mechanism nevertheless falls in line with one of the three courses just described. Only experience can indicate which is the most satisfactory for a given process. Convection, infrared, and direct-fired ovens can be used to achieve fusion, but the necessary precautions should be taken to safeguard the process against the hazards of conventional solvents.

APPLICATIONS OF VARIOUS DISPERSION COATINGS

Considerable leeway exists with respect to the method of application of vinyl dispersion coatings to various substrates, whether they be solid or relatively porous in nature. Production people employ such techniques of applying vinyl paste coatings as knife over blanket, knife over roller, reverse roller, dipping, spray coating, extrusion coating,

and the use of various types of moulding methods such as slush moulding, rotational casting, and cavity moulding.

Reverse Roller Coating

Reverse roller coating is widely accepted as a method for applying
vinyl paste dispersions. Coatings of this nature can be formulated so
that one can apply them to relatively porous substrates by means of the
techniques just described. In reverse roller coating one roll runs in a
direction opposite to that of the substrate to be coated. This feature
also permits considerable leeway in coating weights and high operating speeds.

Knife Coating

In the knife-coating technique, a stationary coating knife is used to
meter the coating as the substrate passes under it. Although the knife
coater is far more simple in basic design than the reverse roll coater,
control of coat thickness is more difficult.

Where the substrate is relatively open or porous, one may need a
prime coat to seal the surface and prevent subsequent coatings from
penetrating into the web. Here, modified reverse coaters or knife coaters can be used with coating media of relatively high viscosity.

Dip Coating

Dip-coating techniques lend themselves to the coating of such materials as wire, cordage, canvas, and similar products. With careful adjustment of the viscosity characteristics, varying the rate at which the
material is dipped, one can control or regulate the thickness of coating
deposited on a suitable substrate by using such mechanical devices as
blades and scrapers. Preheating the substrate sometimes reduces the
number of passes through the equipment and the use of thickening
agents in the formulation enables one to build up heavier coatings and
still prevent sagging or dripping.

Spray Coating

Suitably formulated vinyl paste dispersion coatings can be spray-
coated. In general those containing active volatile solvents offer fewer
difficulties and added advantages accrue if the percentage of nonvolatiles is kept relatively low.

Extrusion Coating

Another method of applying coatings of this nature to substrates
employs extrusion equipment. This procedure differs from both knife

and roller methods in that the coating mixture is fused prior to application to the substrate. The vinyl dispersion coating is pumped into the hot extruder, where it is fused and forced out through a film die. The resulting hot film is then brought into contact with the substrate in the nip of two rolls, where lamination of the hot film and the substrate occurs.

Moulding

Although the use of moulding equipment is not of necessity a method of applying a coating to a substrate, it is nevertheless a method of using vinyl dispersion coatings. Several types of moulding technique are applicable to this type of coating formulation. They are slush moulding, rotational casting, and cavity moulding.

Slush Moulding

Slush moulding is performed by filling a mould with the vinyl coating; the mould is heated to the point where the coating next to the wall of the mould builds up to a specified thickness. The mould is then turned upside down and the coating remaining in liquid form is dumped out and can be used again in a similar manner. In the next operation, the mould with the semifused material next to its walls is then fused in an oven. After fusion occurs the material can then be removed from the mould without distortion or tearing. Preheated moulds are especially useful in quickly building up thick sections.

Rotational Moulding

Another technique at the disposal of the formulator is rotational moulding which is similar to slush moulding. In this instance a vinyl coating is charged to a hollow thin-walled sectional mould, the mould is then closed and allowed to rotate in two or more planes at the same time as it is being heated to gel and fuse the coating. After adequate fusion time the mould is cooled, opened, and the moulded article removed.

Cavity Moulding

Cavity moulds can also be used with vinyl dispersion coatings. This method includes the pumping or pouring of the dispersion into a cavity until that cavity is filled; the mould is then brought to the fusion temperature of the coating. With this technique the cross-sectional thickness of the moulded part determines the fusion time required. Even heavy sections several inches thick can be moulded with comparative ease by means of the cavity method.

FOAM COATINGS

Vinyl dispersion coatings to produce foamed or sponged articles on various substrates can be achieved by the incorporation of chemical blowing agents or by "whipping" a gas or mixture of gases into the vinyl dispersion. The material so produced can be in continuous sheet form or in open or closed moulds depending on the method of application and the end use of the articles to be prepared. Such material forms resilient durable foamed products.

One technique for producing connected cellular-material is by the use of a nitrogen-liberating blowing agent. To produce open-cell foam structure the decomposition of the blowing agent (which releases the nitrogen gas) must be simultaneous with the gellation of the paste dispersion thus rupturing the cell walls and allowing them to be interconnected. With closed-cell foams the coating gel must fuse prior to the decomposition of the blowing agent or release of the gaseous material, thereby trapping the gas. Open-cell foam can be produced with a closed-cell technique by exposing the gelled or fused paste dispersion at higher temperatures or for longer periods of time than those employed for closed-cell material thus causing the walls of the cells to fracture—a feature that is commonly termed "over blowing."

USES OF VINYL DISPERSION COATINGS

To summarize: as outlined in the opening paragraphs of this chapter vinyl dispersion coatings have found numerous end uses and can be readily applied to the paper coating and cloth coating industries where the end product would find use in upholstery, interior automobile trim, wallcovering, and garment materials. The metal decorating industries use this type of coating to produce both protective and decorative coats on various metals. Wire dipping and moulded metal-formed parts also employ coatings of this nature. A wide variety of products such as hollow toys, dolls, shoes, rain boots, and other wearing apparel can be produced by moulding techniques.

Index

www.ingramcontent.com/pod-product-compliance
Lightning Source LLC
Chambersburg PA
CBHW060810220326
41598CB00022B/2578